Lecture Notes in Computer Science 9630

Commenced Publication in 1973
Founding and Former Series Editors:
Gerhard Goos, Juris Hartmanis, and Jan van Leeuwen

More information about this series at http://www.springer.com/series/8851

Ngoc Thanh Nguyen · Ryszard Kowalczyk
Paulo Rupino da Cunha (Eds.)

Transactions on Computational Collective Intelligence XXI

Special Issue on Keyword Search and Big Data

 Springer

Editors-in-Chief
Ngoc Thanh Nguyen
Wrocław University of Technology
Wroclaw
Poland

Ryszard Kowalczyk
Swinburne University of Technology
Hawthorn, VIC
Australia

Guest Editor
Paulo Rupino da Cunha
CISUC, Department of Informatics
  Engineering
Universidade de Coimbra
Coimbra
Portugal

ISSN 0302-9743   ISSN 1611-3349 (electronic)
Lecture Notes in Computer Science
ISBN 978-3-662-49520-9   ISBN 978-3-662-49521-6 (eBook)
DOI 10.1007/978-3-662-49521-6

Library of Congress Control Number: 2016931292

Printed on acid-free paper

This Springer imprint is published by SpringerNature
The registered company is Springer-Verlag GmbH Berlin Heidelberg

Transactions on Computational Collective Intelligence XXI
Special Issue on Keyword Search and Big Data

Preface

Keyword-based search has become the de facto standard for information discovery on the Web. This is mainly due to its simplicity, when compared with more formal query languages such as SQL or SPARQL. Although these are more efficient and effective, they require specialized knowledge usually held by computer scientists and engineers. However, as the Web evolves and new structured databases are brought on-line, as semantics, social, and contextual aspects associated with the data become more relevant, the challenges to the feasibility of keyword-based search increase. But these have to be overcome, if the latent socio economic impact of the new data sources is to be unlocked by end-users. For keyword-based search to be possible in this new reality, various problems have to be addressed. Furthermore, these are multifaceted in nature and the solutions require contributions from many disciplines. This special issue provides a contribution to this research space.

In the first paper, entitled "Keyword-Based Search over Databases: A Roadmap for a Reference Architecture Paired with an Evaluation Framework," by Sonia Bergamaschi, Nicola Ferro, Francesco Guerra, and Gianmaria Silvello, the authors argue the need for a reference architecture for keyword search in databases, as well as a companion evaluation framework. The goal is to focus on the end-user goals, both by considering compelling use cases and by bringing together diverse techniques to fully address them. In the quest for a more effective keyword-based search, Enrico Sartori, Yannis Velegrakis, and Francesco Guerra, the authors of the second paper, "Entity-Based Keyword Search in Web Documents," propose a way to increase the expressive power of a source document by representing it as a set of structures that describe relationships among the entities mentioned in the text. The third paper, "Evaluation of Keyword Search in Affective Multimedia Databases," discusses a case where keyword-based search is used beyond the realms of simple text documents. In this case, the authors – Marko Horvat, Marin Vukovic, and Željka Car – present an experimental evaluation of multi-keyword search in affective multimedia databases, where the emotional responses to the pieces of content are stored with a description of their semantics using keywords from unsupervised glossaries, expected emotion elicitation potential, and other contextual information. Another field of application of keyword-based search is addressed in the fourth paper – "Data-Driven Discovery of Attribute Dictionaries" – where Fei Chiang, Periklis Andritsos, and Renee J. Miller present techniques to extract attribute values from textual product descriptions of on-line retailers. Automated techniques such as this are increasingly important to ensure effective searches by customers, as the number of products available at these outlets increase. Moving on to social media, the next paper addresses the challenge of efficiently retrieving and understanding messages. In "Subject-Related Message Filtering in Social Media Through Context-Enriched Language Models," Alexandre Davis and Adriano Veloso exploit context in the analysis of social media messages using computational linguistics techniques, with interesting results. The importance of context is again emphasized in the sixth paper – "Improving Open Information Extraction for Semantic Web Tasks" – by Cheikh Kacfah Emani, Catarina Ferreira da Silva, Bruno Fiès, and Parisa Ghodous, who take Open Information Extraction, which aims at automatically identifying all the possible assertions within a sentence, and show how it can be improved when working on a given domain of knowledge. Last but not least, in "Searching Web 2.0 Data Through Entity-Based Aggregation," Ekaterini Ioannou and Yannis Velegrakis draw on suggestions from the Linked Data movement to design an approach for more efficient and effective integration of Social and Semantic Web data using entities.

The papers in this special issue have been accepted after, at least, two iterations of a double-blind peer-review process. I would like to express my sincere thanks to all the authors and the Reviewers of the Special Issue for their contributions. A special word of appreciation to the Editor-in-Chief, Ngoc Thanh Nguyen, for embracing this project, and, finally my gratitude to Bernadetta Maleszka, Assistant Editor of TCCI, for all the help throughout the review process and in preparing the final documents for the issue.

January 2016   Paulo Rupino da Cunha

Transactions on Computational Collective Intelligence

This Springer journal focuses on research in applications of the computer-based methods of computational collective intelligence (CCI) and their applications in a wide range of fields such as the Semantic Web, social networks, and multi-agent systems. It aims to provide a forum for the presentation of scientific research and technological achievements accomplished by the international community.

The topics addressed by this journal include all solutions to real-life problems for which it is necessary to use CCI technologies to achieve effective results. The emphasis of the papers published is on novel and original research and technological advancements. Special features on specific topics are welcome.

Editor-in-Chief

Ngoc Thanh Nguyen   Wroclaw University of Technology, Poland

Co-Editor-in-Chief

Ryszard Kowalczyk   Swinburne University of Technology, Australia

Editorial Board

John Breslin	National University of Ireland, Galway, Ireland
Shi-Kuo Chang	University of Pittsburgh, USA
Longbing Cao	University of Technology Sydney, Australia
Oscar Cordon	European Centre for Soft Computing, Spain
Tzung-Pei Hong	National University of Kaohsiung, Taiwan
Gordan Jezic	University of Zagreb, Croatia
Piotr Jędrzejowicz	Gdynia Maritime University, Poland
Kang-Huyn Jo	University of Ulsan, Korea
Jozef Korbicz	University of Zielona Gora, Poland
Hoai An Le Thi	Metz University, France
Pierre Lévy	University of Ottawa, Canada
Tokuro Matsuo	Yamagata University, Japan
Kazumi Nakamatsu	University of Hyogo, Japan
Toyoaki Nishida	Kyoto University, Japan
Manuel Núñez	Universidad Complutense de Madrid, Spain
Julian Padget	University of Bath, UK
Witold Pedrycz	University of Alberta, Canada

Reviewers of the Special Issue

Contents

Keyword-Based Search Over Databases: A Roadmap for a Reference Architecture Paired with an Evaluation Framework

Sonia Bergamaschi[1], Nicola Ferro[2], Francesco Guerra[1(✉)], and Gianmaria Silvello[2]

[1] Department of Engineering "Enzo Ferrari",
University of Modena and Reggio Emilia, Modena, Italy
{sonia.bergamaschi,francesco.guerra}@unimore.it
[2] Department of Information Engineering, University of Padua, Padua, Italy
{ferro,silvello}@dei.unipd.it

Abstract. Structured data sources promise to be the next driver of a significant socio-economic impact for both people and companies. Nevertheless, accessing them through formal languages, such as SQL or SPARQL, can become cumbersome and frustrating for end-users. To overcome this issue, keyword search in databases is becoming the technology of choice, even if it suffers from efficiency and effectiveness problems that prevent it from being adopted at Web scale.

In this paper, we motivate the need for a reference architecture for keyword search in databases to favor the development of scalable and effective components, also borrowing methods from neighbor fields, such as information retrieval and natural language processing. Moreover, we point out the need for a companion evaluation framework, able to assess the efficiency and the effectiveness of such new systems and in the light of real and compelling use cases.

1 Introduction

Since the last decade, we have been observing a continuous increment of structured data available in the Web. In a first stage, structured data was only indirectly available as "embedded" in Web pages. This is the case of data driven Web applications, where user information needs are constrained by pre-defined Web forms which limit the range of the queries that can be performed in favour of a simpler and more intuitive interaction. Based on the user inputs, the Web application retrieves and publishes data extracted from a private (i.e., not directly accessible by external users) source. The collection of all these sources constitutes the so-called Deep or hidden Web, terms denoting its inaccessible nature. Therefore, the Web application constitutes the entry point for selecting and filtering the access to the data which are then available only by manually filling-up Web forms on application-specific search interfaces [42]. Moreover, tables directly inserted in Web pages have been usually adopted for publishing structured data

N.T. Nguyen et al. (Eds.): TCCI XXI, LNCS 9630, pp. 1–20, 2016.
DOI: 10.1007/978-3-662-49521-6_1

on the Web. Several studies [15] tried to estimate the dimension of the information contained in such tables and to develop techniques for their automatic discovering, extraction and re-use as autonomous structured data sources. This legacy represents still an open issue when it comes to unleashing the full potential of such data sources by allowing users to seamlessly access them in a fashion not tied to pre-defined paths.

More recently, data have been recognized as an extremely valuable asset also from the socio-economical point-of-view; the Economist magazine recently wrote that "data is the new raw material of business" and the European Commission stated that "Big data technology and services are expected to grow worldwide to USD 16.9 billion in 2015 at a compound annual growth rate of 40 % – about seven times that of the information and communications technology (ICT) market overall" [22]. The principal driver of this evolution is the Web of Data, the size of which is estimated to have exceeded 100 billion facts (i.e. semantically connected entities). The actual paradigm realizing the Web of Data is the Linked (Open) Data [35], which by exploiting Web technologies allows public data in machine-readable formats to be opened up ready for consumption and re-use. In this emerging scenario, huge quantities of structured data are published on the Web and they are readily available to end-users for direct consumption. Furthermore, advanced services (e.g. Web and mobile applications) are increasingly making use of these data by exploiting the outcomes achieved in the semantic Web and Linked Data research fields.

In both cases described above, the tasks of finding data sources well-suited for specific information needs and selecting relevant data within a given data source are crucial. In the following, we will focus on relational databases which are the key and most widespread structured data sources baking the above mentioned scenarios, but we make our analysis general enough to be used also with other sources.

Keyword search is the foremost approach for information searching and it has been extensively studied in the field of Information Retrieval (IR) [14]. Nevertheless, retrieving information from (unstructured or semi-structured) documents is intrinsically different from querying databases, and consequently this model has left out the structured data sources which are typically accessed through structured queries, e.g. Structured Query Language (SQL) queries over relational databases or SPARQL Protocol and RDF Query Language (SPARQL) queries over Linked Data graphs.

Structured queries are not end-user oriented, given that their formulation is based on a quite complex syntax and requires some knowledge about the structure of the data to be queried. Furthermore, structured queries are issued assuming that a correct specification of the user information need exists and that answers are perfect – i.e. they follow the "exact match" search paradigm. On the other hand, end-users are more oriented towards a "best match" search paradigm given that their information needs are often vague and subjected to a progressive and gradual process of refinement enabled by the search activity itself [7]. Indeed, according to Marchionini [44], information seeking activities can be aimed at lookup (e.g. fact checking), learn (e.g. comprehend a phenomena, developing new knowledge) and investigate (e.g. support planning and forecasting) where learning and investigative searching require strong human involvement in a continuous and interactive way. As a consequence, search can also be seen as a multi-stage activity that helps the user to clarify her/his information needs and to subsequently tune her/his queries for finding better suited results.

In the last fifteen years, these facts triggered the research community to put a lot of effort in developing new approaches for keyword search over structured databases [16,17,53]. Nevertheless, despite of the research work, there are not yet keyword search prototypes able to scale up to industrial-grade applications. Two main issues are hampering the design and development of next generation systems for keyword search over structured data: (i) the lack of systemic approaches considering all of the issues of keyword search from the interpretation of the user needs, to the computation, retrieval, ranking and presentation of the results; and (ii) the absence of a shared and complete evaluation methodology measuring user satisfaction, achieved utility and required effort for carrying out informative tasks.

In particular, with respect to the first issue, we claim that a conceptual architecture pivoting around keyword search and structured data needs to couple system- and user-oriented components. The former ones aim at augmenting the performances (i.e. efficiency) of the search, whereas the latter ones aim at improving the quality of the search (i.e. effectiveness) from the user perspective. Such system- and user-oriented components already exist and have been demonstrated to be effective for their specific purpose. Nevertheless, their integration into a unique framework for keyword search is still lacking. Moreover, it should be considered that a search process is part of a wider user task [37] from which the information need arises and, in turn, makes the user resort to issuing queries to satisfy it. This makes the whole process quite complex and brings in the accomplishment of the user information need several degrees of uncertainty. The main focus of keyword search, i.e. getting out the most from the relational data starting from keywords instead of a structured query, is certainly something that helps users in carrying out their tasks. Nevertheless, many factors, beyond algorithmic correctness and completeness, may impair the impact of keyword search and prevent users to fully exploit its potentialities. Indeed, as also discussed above, even if a database is designed for answering exact queries, the search process and the user intents are usually defined in a vague way which calls for best match approaches and for ranking the results of a query (even exact) by the estimation of how much they may fit the actual information needs of the user. Therefore, we need to put keyword search into a broader context and envision innovative architectures where keyword search is one of the components, paired with other building blocks to better take into account the variability and uncertainty entailed by the whole search process.

Concerning the second issue, experimental evaluation [31] – both laboratory and interactive – estimates how much systems are adherent to the user information needs, provides the desired effectiveness and efficiency, guarantees the required robustness and reliability and operates with the necessary scalability. Measuring these abilities provides insights about the features of a system thus indirectly becoming a key means for supporting and fostering the development of new systems with improved performances.

In light of this, we claim that the current frameworks for the evaluation of keyword search in relational databases [19] need to be re-thought, by moving beyond the evaluation of keyword search components in isolation or not related to the actual user needs, and, instead, by considering the whole system, its constituents, and their inter-relations with the ultimate goal of supporting actual user search tasks [9, 24]. Furthermore, we outline some guidelines for defining a fair and complete evaluation of keyword search approaches building on the well-established IR evaluation methodologies and tuning it for taking into account the intrinsic peculiarities of search over structured data.

The paper is organized as follows: Sect. 2 discusses the common approaches for querying structured data, proposes a conceptual architecture for keyword-based search systems and presents a use case. Section 3 points out the main limits of current benchmarks for evaluating keyword search systems over structured data and Sect. 4 outlines some guidelines for defining a fair and complete evaluation methodology. Finally, Sect. 5 reports final remarks and future works.

2   Querying Structured Data

Issues related to keyword search over documents have been addressed by the IR community since its inception, back in the mid of last century. A generic IR system comprises three main components [14]: an index and a retrieval and ranking model. The typical search process starts by considering the user's *information need* from which a query composed of one or more keywords (i.e. *terms*) is then derived. Although search is typically based on simple keywords, IR systems usually support also richer syntax allowing for the use of boolean and pattern matching operators, compound terms, phrases, alternative weighting of terms, and so on. A major task of an IR system is to build and maintain the *index* which is a data structure containing the terms in the documents associated with the locations where they appear; for enhancing the search process IR systems also maintain statistics about the indexed documents such as the number of occurrences of a term in a document and in the whole collection. By exploiting the index, the statistics and other available data, the system processes the user query (i.e. parse it into terms to be matched against the index) and by using a retrieval model – e.g. boolean model, vector space model, language model – returns a list of documents ranked accordingly to their estimated relevance to the user need. Finally, evaluation is conducted off line and it is aimed at measuring the effectiveness of an IR system for further understanding its functioning in a real environment and giving insights about how it can be improved. As stated in [14], in order to be successfully carry out, IR evaluation must have: (i) a characterization of the system purpose; (ii) a measure that quantifies how well this purpose is met; and, (iii) an accurate and economical measurement technique.

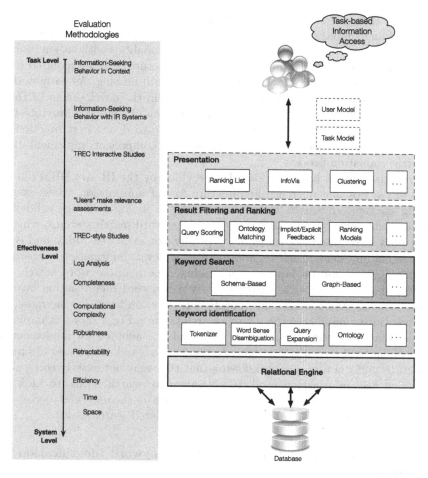

Fig. 1. Reference architecture and evaluation methodologies for a keyword-based search system.

The problem of ranking documents (i.e. assign *scores*) accordingly to the user's query is one of the most studied problems in the field.

IR can be considered now as a mature technology and its research activity outcomes are implemented in commercial systems such as web search engines and domain-specific IR systems (e.g. desktop search, patent search and medical search). Furthermore, IR can count on a well-defined and shared specification of the main components of a search system, a thorough and fair evaluation methodology and a cooperative research community addressing specific aspects of IR systems – e.g. information needs modeling, indexing, retrieval and ranking and results presentation – while maintaining the common aim of improving search efficiency and effectiveness.

The research in keyword search over relational databases is still young and far from providing similar outcomes. In particular, the approaches proposed in the literature typically focus on developing some functionalities without envisioning a holistic solution. A study and a reference architecture defining required functional components and interfaces among them is still missing. As witnessed by the foremost role of the ANSI-SPARC model [50] in the development of Data-Base Management System (DBMS), the specification of a standard architecture can provide several advantages, making it possible to define best practices as well as to enable interoperability among different component implementations and a fair evaluation process.

Leveraging on the base of knowledge provided by the IR and DBMS areas, we envisaged possible reference architecture for an information access system pivoting on keyword search techniques, as shown in Fig. 1. Thick solid lines frame modules which are the focus of current keyword search systems, whereas dotted lines frame modules which are typically not exploited today and come from neighboring fields, such as information retrieval, information extraction, data mining, and natural language processing. The main keyword search and relational database layers are surrounded by a Keyword Identification layer, a Results Filtering and Ranking layer, and a Presentation layer. Furthermore, for each constituting module of the reference architecture, Fig. 1 shows examples of possible components and technologies that could be adopted for implementing it. Indeed, an ideal search system has to consider the search task a user desires to conduct, to perform user queries knowing that they may not exactly correspond to the real user information need, to disambiguate search terms, to rank the results of the search process on the basis of relevance for the user, and to visualize these results in the most proper way for the considered search task.

2.1 Understanding the User Input: The Keyword Identification Module

The final goal of the *Keyword Identification* layer is to understand what the user had in mind when s/he formulated the query. This task requires the development of techniques for understanding the data structures involved in the query and their correlations. Such layer is not required in the current DBMSs which are typically queried by a structured query language, e.g. SQL for relational DB. In these systems, the user is in charge of specifying in his/her query which are the data structures containing the data of interest, and how these structures are linked with each other for the computation of an answer [10]. This way, users can formulate accurate and exact queries. Nevertheless, the expressive power of the structured query languages comes with a price: it requires users to be expert, since they have to know the syntax of the language and the data source structures. This limitation makes this kind of language not suitable for users who do not know the language adopted or the source structures storing the data.

A querying system based on keywords does not require the knowledge of any language and data structure, thus allowing people to formulate queries in an easier way. Nevertheless, keyword queries are inherently ambiguous: the same keyword may refer to terms belonging to different database structures. This problem is also common to IR systems, where similarly the same keyword may be used with different meanings in the same/different document(s). In relational databases, this ambiguity generates more issues, mainly due to the fragmentation of the information in different tables. In case of multiple keywords, for example, we have not only to deal with the uncertainty about the meaning of each term, but also about the paths connecting the selected data structures to be followed. The choice of one path instead of another can generate different results since they are based on different interpretations of the intended meaning of the keyword query (see Sect. 2.4).

Recently, some formal keyword-based languages have been proposed to combine simplicity and intuitiveness of expression with the richness of more expressive languages, such as the Contextual Query Language (CQL). Nevertheless, these languages are not commonly adopted and their expressiveness is still low compared what the SQL language allows users to formulate in databases.

The last difference between IR systems and keyword search over structured databases is that in IR user keywords are directly retrieved in some specific indexes over the data, whereas in the so-called schema-based keyword search systems the keywords provide the information for formulating structured queries, but they are not directly used by the system for retrieving data.

It is evident that the Keyword Identification layer relies on different components for managing the different approaches.

Summarizing, the Keyword Identification is a strategic layer in keyword search over structured database, since it is in charge of understanding what the user had in mind when formulating the query.

The final goal is freeing keyword search from an exact match with the keywords present in the relational data and introducing the possibility of matching in multiple ways the keywords expressed by user in order to compensate for possible imprecisions or errors in the choice of the keywords.

2.2 The Business Logic: Keyword Search, Result Filter, and Rank

The *Keyword search* layer aims at matching the user keywords (or their interpretations) with data structures and domains of selected attributes. Keyword search aims at retrieving the database tuples matching the user keywords (or their interpretations). Two main techniques are typically adopted [53]: graph-based and schema-based. Graph-based techniques (e.g., BANKS [1], BLINKS [33], PRECIS [47], DPBF [21] and STAR [39]) model relational databases as graphs, where nodes are tuples, edges foreign-primary key relationships between those tuples. Their main aim is to optimize the computation of specific structures over the graphs (e.g., Steiner trees, rooted trees, etc.) to find the most relevant top-k connected tuples. Their challenge is to handle the large and complex graphs induced by the database instance, as it could make the problem hardly tractable. Schema-based techniques (e.g., DISCOVER [36], DBXplorer [4], SPARK [43], and SQAK [49]) exploit the schema information to formulate SQL queries determined starting from the user keyword queries. In this case, the system has to discover the structures containing the keywords and how these structures may be joined in order to formulate a set of queries capturing the intended meaning of the user keyword query, expressed in the native structured query language of the source. Most of the existing approaches rely on indexes and functions over the data values to select the most prominent tuples as results of queries.

Recently metadata-based approaches have been developed [8,12,13]. These approaches are useful when there is not direct access to the database instance or when frequent updates make the process of building and updating indexes too expensive.

Finally, some hybrid approaches, combining the features of graph and schema-based systems have been proposed. Among them, QUEST [11] couples by means of a probabilistic framework a schema-based approach for matching keywords with data structures and domains (thus retrieving solutions according to the "user perspective", i.e., to what the user had in mind when s/he formulated the query), and a graph-based approach to connect the tables that better represent the meaning of the user query (thus retrieving solutions according to the "database perspective", i.e., to the way data is stored in the database).

The *Results Filtering and Ranking* layer accounts for the need of adopting alternative strategies for ranking and selecting the results to be presented to the user by the system. This layer plays a fundamental role when the "exact match" search paradigm is left in favor of a "best match" approach where the results returned to the user are not always "correct", but need to be ranked on the basis of the relevance to the user information need. It may concern weighting the results on the basis of the process which generated relational queries from user keywords, or relying on implicit/explicit feedback from the user to filter out some results, or using rank aggregation and data fusion techniques to merge alternative ranking strategies. In some of the approaches this layer is embedded and joint with the keyword search layer.

Fig. 2. A fraction of a database schema with its data.

2.3   Presentation of the Results

The *Presentation* layer regards how the outputs of a system are presented to the user; for instance, we can have traditional ranking lists, results presentation based on advanced information visualization techniques and user interfaces [34], presentations of clusters of results.

The presentation, conversely to what happens in IR, is not standardized. In IR, it is clear that users are querying over a set of documents and they are expecting as output a ranked list of them on the basis of their information needs. In databases, most of the keyword search approaches work over a single database and not over a set of databases (the granularity of the result is different). In this case, it is not clear which answer should be returned to the user. It could be a boolean value (i.e., the database contains the user keywords), or tuples with different level of granularity (from few columns to the universal tuple). The definition of which is the best "dimension" of the results for a specific query is still an open challenge in current systems.

2.4   Use Case

To better illustrate the need for a reference architecture, let us introduce a simple relational database containing tourism information about people, events planned in some cities and possible accommodations for tourists. In Fig. 2, we report a fraction of the database schema with some sample data. Some descriptive attributes are defined for each table and some foreign key - primary key relationship (represented with dashed lines) exists between the tables.

As mentioned, some of the issues addressed by keyword search systems in databases are common to IR systems searching in documents. The process for tokenizing keywords, or recognizing synonym terms, for example, as most of the components included in the Keyword Identification layer in Fig. 1 can indeed be the same in both the scenarios given that the approaches share the same goal: satisfying the user information need. Nevertheless, relational databases differ from documents because they comprise a schema. This has two major implications: (i) the schema can provide useful information for solving the query; and (ii) the schema elements can be used for augmenting the search space given that keywords can match also with schema elements (i.e. metadata) and not only with the data contained in the database.

In the following, we consider queries composed of two keywords that we assume to be conjunct by a logical AND operator; we can point-out three possible matches between the keywords and databases:

– *data-oriented matches*: both keywords match into data instances;
– *schema-oriented matches*: both keywords match into schema elements;
– *mixed matches*: a keyword matches into a schema element, the other into a data instance.

Data-Oriented Matches. Let us consider the keyword query "Spoleto Charleston": the process for retrieving its possible answers can be schematized in two main steps.

In the *first step*, an index-based approach can find several matches for the keyword "Spoleto" with instances of the table Accommodation (i.e., Spoleto is the name of a Hotel, or Spoleto is a value of the attribute City in the same table) with an instance of the table Event (i.e. Spoleto is contained in the title of an event) or with an instance of the table City (i.e. Spoleto is the name of the city). This is due to the fact that the user query is ambiguous and it is unclear whether the user is looking for hotels called "Spoleto", hotels which are in "Spoleto", or the event "Spoleto Festival".

All the three answers are valid and possibly relevant to the user that issued such a query; indeed, one of the tasks of the search engine is to rank them by the estimated relevance to the user information need. The very same scenario can be outlined by considering the second keyword – i.e. "Charleston" – which can match with an element of the table City or of the table Event.

In the *second step*, the possible paths – that need to be formalized into structured queries when working with schema-based approaches – connecting the elements identified in the previous step have to be computed and ranked. For example, if we consider a keyword query "John Charleston", we can individuate two possible paths linking the tables People and City: one through the table Living, meaning that the user is interested in some people called "John" living in "Charleston" (see Fig. 3(a)) and the second through the tables Participating and Event, meaning that the user is interested in some people called "John" participating in an event organized in "Charleston" (see Fig. 3(b)). This problem can be even more complex if some keywords match with foreign key - primary key values. This occurs in our example if the user formulates the query "Spoleto", where it is unspecified if the user is interested in finding people partecipating in an event called "Spoleto Festival" (i.e., "Spoleto" is an instance of Participating), in the event itself (i.e., "Spoleto" is an instance of Event) or in the city (i.e., "Spoleto" is an instance of City). From an algorithmic perspective an answer showing all partecipants in the Spoleto Festival provides an answer as good as the one that reports all the information about the event or the city. Nevertheless, from the user perspective, only one of these sets of results is relevant to her/his information needs, e.g. the city one, while providing all of them would reduce the effectiveness of the system and hampers performance.

Finally, note that our assumption that keywords are connected with AND operators can have nontrivial implications in databases, in particular when both keywords match with the same attribute domain. One of the possible answers of the query *"Spoleto Charleston"* matches both keywords into elements of the table City. In this case, the simplest interpretation of the query – i.e. "the user is looking for cities called Spoleto and Charleston" – would return no results because of the first Normal Form in databases which imposes atomic values in the attribute domains. Possible not empty interpretations of the query are "people living in Spoleto and Charleston", or "people participating in events organized in

(a) First path: John living in Charleston

(b) Second path: John partecipating in an event in Charleston

Fig. 3. Possible paths solving a keyword query

Spoleto and in Charleston" that requires, in schema-based approaches, the non trivial operation of understanding it and the consequent formulation of complex join SQL queries involving other tables.

Schema-Oriented Matches. Several existing keyword search approaches do not consider database metadata as possible targets for user queries and consequently they cannot solve schema-oriented keywords. In these cases, the simple query "Hotel Price" does not find any result. Otherwise, a simple index built on the schema elements can provide the results given that the table Accommodation has the attributes Hotel and Price.

Mixed Matches. Let us consider the query "Hotel Venice", where one keyword refers to a schema element and the second to a data element. Firstly, an index-based approach can find matches between the keyword "Venice" and several possible instances: it could match occurrences of the table Accommodation (i.e., Venice is the name of a Hotel, or Venice is a value of the attribute City), of the table Event (i.e. Venice is the title of an event), and of the table City (i.e. Venice is the name of the city). The user query is ambiguous and it is unclear whether the user is looking for hotels called "Venice", hotels which are in "Venice", or hotel related in some way to the event "Venice Festival". All the listed answers are possible, it is one of the tasks of the search engine to rank them. Moreover, if an index built on the schema element exists (i.e., the system considers database metadata as possible targets for user queries), the second keyword "Hotel" can be exploited to rank the answers. For example, some heuristic rule can prioritize the associations of keywords in the same tables (according with the idea that a query search for something that is "close" in the database representation) and consequently provide higher rank to the results where Venice is an accommodation than the ones where Venice is an event. Note that some keywords can match with schema and data elements at the same time, as it occurs for the keyword "City". In this case, multiple combinations are possible.

Since users do not know the information in the data source, this situation frequently occurs and is typically not managed by most of the existing systems.

A classical keyword search engine may not find any correspondences between the keyword "Hotel" and an element in the database. In this case, the problem is two-fold: firstly, the chosen keyword may better match metadata instead of a value in a table (as in the case in Figure) – indeed, several existing keyword search engines do not consider database metadata as possible targets for user queries; secondly, analogously to what happens in IR systems, the chosen keyword may actually refer to a concept represented in the table with a value which is a synonym of the chosen keyword.

Even in this toy example, it becomes clear that query ambiguity as well as the choice between graph or schema based techniques impacts the system performance. Therefore, the complexity of real scenarios may take even more advantage from the architecture proposed in Fig. 1, which complements keyword search with additional components that, in this specific example, analyze the user keywords and disambiguate their meaning. Similarly, the evaluation methodology must be able to detect these different issues in order to properly assess how systems tackle them.

Finally, as a further example, it is not clear from any of the queries proposed which is the information that the user would like to receive as a result. Several options are possible: in some cases the user would like to receive all the data about the tuples satisfying the criteria defined by the keyword query (i.e., the "universal relation"), in other cases only the values of some attributes (i.e., a projection of the "universal relation"), or even a boolean value (i.e., the existence of at least an instance). On the other hand, she/he also would like to receive results ordered by the system estimation of their relevance and how much they satisfy her/his information need. The task of understanding the granularity, the ordering and aggregation of the expected results is not typically managed by the existing systems and requires a specific module in the architecture of a complete keyword search system. As above, evaluation must be able to take into account these aspects and assess the systems accordingly.

3   Evaluating Keyword-Based Search Approaches for Structured Data Sources

Innovative proposals for pushing the boundaries of keyword search cannot set aside a proper and shared evaluation methodology which helps in progressing towards the envisioned goals, ensures the soundness and quality of the proposed solutions, and guarantees the repeatability and comparability of the experiments.

As shown on the left hand side of Fig. 1, evaluation can be carried out at three levels: at a "task level" for instance by means of user studies [41]; at an "effectiveness level" by means of the test collection methodology [31]; and, at a "system level" by means of benchmarking queries per second, memory and CPU load, correctness and completeness [19], thus progressively moving from a human to a system focus.

Nevertheless, experimental evaluation is hampered by fragmentation – different tasks, different collections, different perspectives from interactive to laboratory evaluation which are usually dealt with in separated ways, without sharing resources and the produced data [52]. This will be even more true for the multidisciplinary approach to the system architecture proposed in Fig. 1; for this reason, a unified and holistic approach to evaluation would be needed to assess the different facts of such a complex system and to reconcile the experimental outcomes.

To the best of our knowledge, only few papers addressing issues related to the evaluation of keyword search systems over relational databases have been published in the literature. In [51], it is observed that the existing keyword search approaches have been evaluated against different databases with different query sets. This fact prevents their direct comparison based on their original experimental results, since cross-collection comparisons are often not feasible or, at least, extremely difficult. Moreover, in some cases, the evaluation framework adopted appears to be inadequate, mainly due to the employment of a small number of self-authored queries, thus leading to biased results. Only recently a benchmark [19] proposed some metrics and a query set for evaluating the approaches against three data sources (Mondial, IMDB and Wikipedia). Even if it represents an important step towards a fair evaluation of keyword search approaches, the benchmark suffers from some limitations. Firstly, the metrics adopted (precision and recall compared to a gold standard, and time needed for returning the results) cannot be suitable when applied to a schema-based keyword search system, which transforms keyword queries into SQL queries. In this case, the evaluation of the time performance is biased by the time required for the execution of the SQL queries by the DBMS underlying the application. Different underlying DBMSs can largely influence the time performance, but they are typically independent of the keyword search technique under evaluation. Secondly, the benchmark computes the effectiveness of the approaches by analyzing the results (instances) retrieved with specific keyword queries whereas schema-based search approaches provide SQL queries as a primary result. Note that all the tuples resulting from the same SQL query have intrinsically the same score, and that the same result can be obtained by different queries. Thirdly, most of the queries in the dataset are composed of only one element. In several approaches the evaluation of this kind of queries is performed only by analyzing the indexes over the data implemented in the DBMS. Even in this case, the evaluation task risks to evaluate the DBMS storing the data instead of the keyword search technique. Fourthly, the benchmark does not discuss what is a "correct" result in terms of granularity. It would be useful to discuss if the system has to retrieve the "universal tuple" or a subset of it.

4   Torwards a Holistic Evaluation Approach

A systematic and comparable experimental evaluation is a very demanding activity, in terms of both time and effort needed to perform it. For this reason, in the IR field, it is usually carried out in publicly open and large-scale evaluation campaigns at international level, which allow for sharing the effort, producing large experimental collections, and comparing state-of-the-art systems and algorithms. Relevant and long-lived examples are Text REtrieval Conference (TREC)[1] [32] in the United States, the Conference and Labs of the Evaluation Forum (CLEF) initiative[2] [25] in Europe, and NII Testbeds and Community for Information access Research (NTCIR)[3] in Japan and Asia. All these international evaluation activities rely on the Cranfield methodology [18] – i.e. the *de-facto* standard for experimental evaluation of IR systems – which makes use of shared experimental collections expressed as a triple composed by a *dataset*, a set of *topics*, which simulates actual user information needs and the *ground-truth* or the set relevance judgments, i.e. a kind of "correct" answers, where for each topic the data relevant for the topic are determined.

This well-established methodology and these international experiences would help in moving evaluation of keyword search in databases forward and fostering the development of next generation systems in this field. Indeed, by sharing resources and providing open fora to compare and discuss approaches, they ease the design of shared evaluation tasks, geared into concrete and compelling use cases as those discussed in the previous sections, which drive the design and development of next-generation systems [2]. Evaluation campaigns support the systematic exploration and deep understanding of the system behaviour via failure analysis [5,29,30], especially when systems are challenged by realistic datasets and compelling use cases. Moreover, they stimulate the creation of multidisciplinary communities, embracing all the competencies needed to embody the reference architecture of Fig. 1. Finally, they foster the re-use of the experimental data and the acquired knowledge [3], giving the possibility to conduct longitudinal studies to track the improvement on performances [6,26], to reproduce the obtained results [27] and lowering the barriers for comparing to and progressing beyond the state-of-the-art [20].

Moreover, as reported by [45], for every $1 that NIST and its partners invested in TREC, at least $3.35 to $5.07 in benefits accrued to researchers and industry. During their life-span, large-scale evaluation campaigns have produced huge amounts of scientific data which are extremely valuable for research and development but also from an economic point of view: [45] estimates that

---

[1] http://trec.nist.gov/.

[2] http://www.clef-initiative.eu/.

[3] http://research.nii.ac.jp/ntcir/.

the overall investment in TREC of about 30 million dollars in its first 20 years which, as discussed above, produced an estimated return on investment between 90 and 150 million dollars. Therefore, applying experimental evaluation to vision represented in Fig. 1 gives the promise not only to advance state-of-the-art techniques, but also to have a concrete economic impact.

Therefore, the evaluation of keyword search approaches assumes a paramount importance. In Sect. 3, some criticisms of the current evaluation frameworks have been highlighted. In the following, we try to go beyond them, by introducing and discussing some guidelines for the definition of an environment for experimenting the keyword search proposals.

Let us focus on scenarios where the search engine is coupled with a single database: this is not there are only few proposals in literature of systems that rank databases on the basis of user keywords).

First of all, it is necessary to define the kind of evaluation we need to conduct because there is a clear distinction between evaluating the *formal correctness* of an algorithm implemented by a system and its *effectiveness* in addressing a search task. In the former case we are evaluating the formal *correctness* of the results returned by an algorithm or a system. The evaluation process assumes the existence of exact results and checks if, given certain inputs (e.g. a query), the actual outputs (e.g. a set of tuples) are the expected ones. Most of the evaluation conducted on keyword-based systems working over structured data has been based on the correctness of the results: in this case, the main difference between algorithms is given by their execution times (i.e. their efficiency) or whether they complete at all, during to their computational complexity. In the latter case, we are considering a *functional* and *holistic* evaluation process that checks if the system being tested is well-suited for addressing a given search task. In this case, we cannot assume the existence of formal correct answers and we are not evaluating the correctness of the implementation of the employed algorithms, but their effectiveness with respect to the search task the system has to address. This switch from correctness to effectiveness should also be accompanied by the adoption of proper evaluation measures [46], capable of grasping different aspects such as the utility delivered to the user [38], the impact of the time the user spends in interacting with the system [48], the effort required to the user [28], or the different ways in which the user scans the result list [23].

Overall, we believe that this evaluation process can lead to substantial improvement of keyword systems over structured data as it happened for document-oriented IR systems. Therefore, we propose an evaluation framework based on the Cranfield methodology which has to be tailored for keyword search over structured data; in the following, we outline the main components (dataset, topics, ground-truth) of this methodology and the issues to be faced.

The Dataset. In general, the dataset has to be representative of the domain of interest both in terms of kinds of data and size. For example, if we consider the case cultural heritage search we need to use actual cultural heritage data as provided by The Library of Congress or Europeana and their size should be close to the actual one – e.g. for search tasks over Europeana, a dataset size of about 100 GB is considered a reasonable choice. This means that the kind of data and its size cannot be fixed a-priori, but have to be decided on the basis of the search task the system has to address.

From the keyword search point-of-view, the dataset must also have a complex structure made of inter-connected tables, with multiple paths joining the same tables, because retrieving data from flat datasets is mainly an index-based process and data indexes are provided by the DBMS. Thus, using a flat dataset would lead to the evaluation of the DBMS and not of the algorithm performing keyword search.

The Topics. One of the main limitations of the current evaluation frameworks is that they propose a set of specific queries to be solved and not a set of information needs from which to derive the queries. This could make the system under evaluation prone to "overfitting" – i.e. the developers may try to adapt their systems with the main goal of optimizing the systems for the specific benchmark queries and not for addressing a general task. In this way, we may evaluate a system specifically tailored on the benchmark, which will not perform accordingly on a different data and query set.

Our idea is to base the evaluation on a set of information needs (i.e. the topics) in the form of short descriptions of what a user is looking for. The set of topics simulates actual user information needs and they could be prepared from real system logs, gathered by means of task-based analysis, or through a deep interaction with the involved stakeholders. As a consequence, the evaluation is conducted starting from the information need and not from ready-to-use query. For instance, a possible information need behind query "Hotel Venice" described in Sect. 2.4 could be "which are the hotels in Venice?".

Moreover, it is necessary to define information needs that can be translated into queries composed of more than one keyword because queries composed of only one term do not test the business logic behind the keyword search system, but only the data indexes, which typically are provided by the DBMS. With reference to the use case presented above, a query like "Venice" requires only to look up a term in a database index. A different situation occurs when keywords match tables connected by multiple paths, where it is not clear which one among all possible paths is the best match for the user information need.

The Ground-Truth. When we evaluate the efficacy of a system, the correctness of the returned results is measured in terms of relevance with respect to user information need. For this reason the possible results to a query have to be judged by a pool of domain users that decide if a result is *relevant* for a given information need – i.e. we have to form the set relevance judgments or the ground-truth against which the system output is evaluated. Note that the relevance judgments can be binary, i.e., relevant or not relevant, or multi-graded, e.g., highly relevant, partially relevant, not relevant and so on [40].

For keyword search systems, the ground-truth creation activity is related to the definition of the expected results from a user keyword query.

A schema-based approach generates (a set of) SQL queries, and they could be a good candidate for the evaluation. Nevertheless, to consider SQL queries as result of a search system makes graph-based approaches (that directly retrieve the data) not comparable with schema-based approaches. Vice versa, even though we consider the tuples returned by the queries, the comparison with graph-based approaches is not straightforward since all the values resulting from the same SQL query have the same rank. For example, the "Hotel Venice" query in schema-based approaches provides a list of hotels, all with the same rank. On the contrary, in graph-based approaches, in principle, each hotel in the list can have a different rank.

However, if we consider tuples as the natural result of user queries we have to define its boundaries. For instance, let us consider the query "Hotel Venice" of the use case, where the user is looking for the hotels located in Venice. A possible result is constituted only by the names of the Hotels (the values of the attribute Hotel in table Accommodation). Another valid result is the entire tuple composed of the values for the attributes Hotel and City. Other results, involving a different number of tables until the reaching of the universal relation, are possible (for example we can join the values of table Accommodation with the ones of City which is connected via foreign key).

5 Conclusion and Future Work

In this paper we discussed the need for considering keyword search over relational databases in the light of broader systems, where keyword search is just one of the components and which are aimed at better supporting users in their search tasks. These more complex systems call for appropriate evaluation methodologies which go beyond what is typically done today, i.e. measuring performances of components mostly in isolation or not related to the actual user needs, and, instead, able to consider the system as a whole, its constituent components, and their inter-relations with the ultimate goal of supporting actual user search tasks.

Future work will be devoted to develop a benchmark that follows the guidelines proposed in the paper. The benchmark will be experimented within the KEYSTONE COST Action[4], which is a network of researchers from 28 countries that aims to study topics related to keyword search in structured databases, and in the context of the CLEF initiative, which is the European forum for the evaluation of information access systems with an emphasis on multilingual and multimodal information with various levels of structure.

References

1. Aditya, B., Bhalotia, G., Chakrabarti, S., Hulgeri, A., Nakhe, C., Parag, P., Sudarshan, S.: BANKS: browsing and keyword searching in relational databases. In: VLDB, Proceedings of 28th International Conference on Very Large Data Bases, Hong Kong, China, 20–23 August 2002, pp. 1083–1086 (2002)

---

[4] http://www.keystone-cost.eu/.

2. Agosti, M., Berendsen, R., Bogers, T., Braschler, M., Buitelaar, P., Choukri, K., Di Nunzio, G.M., Ferro, N., Forner, P., Hanbury, A., Friberg Heppin, K., Hansen, P., Järvelin, A., Larsen, B., Lupu, M., Masiero, I., Müller, H., Peruzzo, S., Petras, V., Piroi, F., de Rijke, M., Santucci, G., Silvello, G., Toms, E.: PROMISE retreat report - prospects and opportunities for information access evaluation. SIGIR Forum **46**(2), 60–84 (2012)

3. Agosti, M., Ferro, N., Thanos, C.: DESIRE 2011: first international workshop on data infrastructures for supporting information retrieval evaluation. In: Proceedings of the 20th International Conference on Information and Knowledge Management (CIKM), pp. 2631–2632. ACM, New York, USA (2011)

4. Agrawal, S., Chaudhuri, S., Das, G.: Dbxplorer: a system for keyword-based search over relational databases. In: Proceedings of the 18th International Conference on Data Engineering, San Jose, CA, USA, 26 February–1 March 2002, pp. 5–16 (2002)

5. Angelini, M., Ferro, N., Santucci, G., Silvello, G.: VIRTUE: a visual tool for information retrieval performance evaluation and failure analysis. J. Vis. Lang. Comput. (JVLC) **25**(4), 394–413 (2014)

6. Armstrong, T.G., Moffat, A., Webber, W., Zobel, J.: Improvements that don't add: ad-hoc retrieval results since 1998. In: Proceedings of the 18th International Conference on Information and Knowledge Management (CIKM 2009), pp. 601–610. ACM, New York (1998)

7. Belkin, N.J., Oddy, R., Brooks, H.M.: SK for information retrieval: part I. Background and theory. J. Documentation **38**(2), 61–71 (1982)

8. Bergamaschi, S., Domnori, E., Guerra, F., Trillo-Lado, R., Velegrakis, Y.: Keyword search over relational databases: a metadata approach. In: Proceedings of the ACM SIGMOD International Conference on Management of Data, SIGMOD, Athens, Greece, 12–16 June 2011, pp. 565–576 (2011)

9. Bergamaschi, S., Ferro, N., Guerra, F., Silvello, G., Search, K.: Evaluation over relational databases: an outlook to the future. In: Proceedings of the 7th International Workshop on Ranking in Databases (DBRank) with VLDB, pp. 8:1–8:3 (2013)

10. Bergamaschi, S., Gelati, G., Guerra, F., Vincini, M.: An intelligent data integration approach for collaborative project management in virtual enterprises. World Wide Web **9**(1), 35–61 (2006)

11. Bergamaschi, S., Guerra, F., Interlandi, M., Trillo-Lado, R., Velegrakis, Y.: QUEST: a keyword search system for relational data based on semantic and machine learning techniques. PVLDB **6**(12), 1222–1225 (2013)

12. Bergamaschi, S., Guerra, F., Rota, S., Velegrakis, Y.: A hidden markov model approach to keyword-based search over relational databases. In: Proceedings of the 30th International Conference Conceptual Modeling - ER, Brussels, Belgium, 31 October–3 November 2011, pp. 411–420 (2011)

13. Blunschi, L., Jossen, C., Kossmann, D., Mori, M., Stockinger, K.: SODA: generating SQL for business users. PVLDB **5**(10), 932–943 (2012)

14. Buettcher, S., Clarke, C.L.A., Cormack, G.V.: Information Retrieval: Implementing and Evaluating Search Engines. The MIT Press, Cambridge (2010)

15. Cafarella, M.J., Halevy, A.Y., Madhavan, J.: Structured data on the web. Commun. ACM **54**(2), 72–79 (2011)

16. Chaudhuri, S., Das, G.: Keyword querying and ranking in databases. PVLDB **2**(2), 1658–1659 (2009)

17. Chu, E., Baid, A., Chai, X., Doan, A., Naughton, J.F.: Combining keyword search and forms for ad hoc querying of databases. In: Proceedings of the ACM SIGMOD International Conference on Management of Data, SIGMOD, Providence, Rhode Island, USA, 29 June–2 July 2009, pp. 349–360 (2009)
18. Cleverdon, C.W.: The cranfield tests on index languages devices. In: Spärck Jones, K., Willett, P. (eds.) Readings in Information Retrieval, pp. 47–60. Morgan Kaufmann Publisher Inc., San Francisco (1997)
19. Coffman, J., Weaver, A.C.: An empirical performance evaluation of relational keyword search techniques. IEEE Trans. Knowl. Data Eng. **26**(1), 30–42 (2014)
20. Di Buccio, E., Di Nunzio, G.M., Ferro, N., Harman, D.K., Maistro, M., Silvello, G.: Unfolding off-the-shelf IR systems for reproducibility. In: Proceedings of SIGIR Workshop on Reproducibility, Inexplicability, and Generalizability of Results (RIGOR) (2015)
21. Ding, B., Yu, J.X., Wang, S., Qin, L., Zhang, X., Lin, X.: Finding top-k min-cost connected trees in databases, pp. 836–845 (2007)
22. European Commission. Communication from the commission to the European parliament, the council, the European economic and social committee and the committee of the regions - towards a thriving data-driven economy. COM (2014). 442 final, http://eur-lex.europa.eu/legal-content/EN/TXT/PDF/?uri=CELEX:52014DC0442&from=EN
23. Ferrante, M., Ferro, N., Maistro, M.: Injecting user models and time into precision via markov chains. In: Proceedings of the 37th Annual International ACM SIGIR Conference on Research and Development in Information Retrieval (SIGIR), pp. 597–606. ACM, New York (2014)
24. Ferro, N. (ed.): Bridging Between Information Retrieval and Databases - PROMISE Winter School, Revised Tutorial Lectures. Lecture Notes in Computer Science (LNCS), vol. 8173. Springer, Heidelberg (2013)
25. Ferro, N.: CLEF 15th birthday: past, present, and future. SIGIR Forum **48**(2), 31–55 (2014)
26. Ferro, N., Silvello, G.: CLEF 15th birthday: what can we learn from ad hoc retrieval? In: Kanoulas, E., Lupu, M., Clough, P., Sanderson, M., Hall, M., Hanbury, A., Toms, E. (eds.) CLEF 2014. LNCS, vol. 8685, pp. 31–43. Springer, Heidelberg (2014)
27. Ferro, N., Silvello, G.: Rank-biased precision reloaded: reproducibility and generalization. In: Hanbury, A., Kazai, G., Rauber, A., Fuhr, N. (eds.) ECIR 2015. LNCS, vol. 9022, pp. 768–780. Springer, Heidelberg (2015)
28. Ferro, N., Silvello, G., Keskustalo, H., Pirkola, A., Järvelin, K.: The twist measure for IR evaluation: taking user's effort into account. J. Am. Soc. Inf. Sci. Technol. (JASIST) (in print)
29. Harman, D., Buckley, C.: SIGIR 2004 workshop: RIA and "Where can IR go from here?". ACM SIGIR Forum **38**(2), 45–49 (2004)
30. Harman, D., Buckley, C.: Overview of the reliable information access workshop. Inf. Retrieval **12**(6), 615–641 (2009)
31. Harman, D.K.: Information Retrieval Evaluation. Morgan and Claypool Publishers, USA (2011)
32. Harman, D.K., Voorhees, E.M. (eds.): TREC. Experiment and Evaluation in Information Retrieval. MIT Press, Cambridge (2005)
33. He, H., Wang, H., Yang, J., Yu, P.S.: BLINKS: ranked keyword searches on graphs. In: Proceedings of the ACM SIGMOD International Conference on Management of Data, Beijing, China, 12–14 June 2007, pp. 305–316 (2007)

34. Hearst, M.A.: Search User Interfaces, 1st edn. Cambridge University Press, New York (2009)
35. Heath, T., Bizer, C.: Linked Data: Evolving the Web into a Global Data Space. Synthesis Lectures on the Semantic Web. Morgan and Claypool Publishers, USA (2011)
36. Hristidis, V., Papakonstantinou, Y.: DISCOVER: keyword search in relational databases. In: VLDB, Proceedings of 28th International Conference on Very Large Data Bases, 20–23 August 2002, Hong Kong, China, pp. 670–681 (2002)
37. Ingwersen, P., Järvelin, K.: The Turn: Integration of Information Seeking and Retrieval in Context. Springer, Heidelberg (2005)
38. Järvelin, K., Kekäläinen, J.: Cumulated gain-based evaluation of IR techniques. ACM Trans. Inf. Syst. (TOIS) **20**(4), 422–446 (2002)
39. Kasneci, G., Ramanath, M., Sozio, M., Suchanek, F.M., Weikum, G.: STAR: steiner-tree approximation in relationship graphs. In: Proceedings of the 25th International Conference on Data Engineering, pp. 868–879. IEEE Computer Society (2009)
40. Kekäläinen, J., Järvelin, K.: Using graded relevance assessments in IR evaluation. J. Am. Soc. Inf. Sci. Technol. (JASIST) **53**(13), 1120–1129 (2002)
41. Kelly, D.: Methods for evaluating interactive information retrieval systems with users. Found. Trends Inf. Retrieval (FnTIR) **3**(1–2), 1–224 (2009)
42. Khare, R., An, Y., Song, I.-Y.: Understanding deep web search interfaces: a survey. SIGMOD Rec. **39**(1), 33–40 (2010)
43. Luo, Y., Lin, X., Wang, W., Zhou, X.: SPARK: top-k keyword query in relational databases. In: Proceedings of ACM SIGMOD International Conference on Management Of Data (SIGMOD), pp. 115–126. ACM, New York (2007)
44. Marchionini, G.: Exploratory search: from finding to understanding. Commun. ACM **49**(4), 41–46 (2006)
45. Rowe, B.R., Wood, D.W., Link, A.L., Simoni, D.A.: Economic impact assessment of NIST's text retrieval conference (TREC) program. RTI Project Number 0211875, RTI International, USA, July 2010. http://trec.nist.gov/pubs/2010.economic.impact.pdf
46. Sakai, T.: Metrics, statistics, tests. In: Ferro, N. (ed.) PROMISE Winter School 2013. LNCS, vol. 8173, pp. 116–163. Springer, Heidelberg (2014)
47. Simitsis, A., Koutrika, G., Ioannidis, Y.E.: Précis: from unstructured keywords as queries to structured databases as answers. VLDB J. **17**(1), 117–149 (2008)
48. Smucker, M.D., Clarke, C.L.A.: Time-based calibration of effectiveness measures. In: Proceedings of the 35th Annual International ACM SIGIR Conference on Research and Development in Information Retrieval (SIGIR), pp. 95–104. ACM, New York (2012)
49. Tata, S., Lohman, G.M.: SQAK: doing more with keywords. In: Proceedings of ACM SIGMOD International Conference on Management of Data (SIGMOD 2008), pp. 889–902. ACM Press, New York (2014)
50. Tsichritzis, D., Klug, A.: The ANSI/X3/SPARC DBMS framework report of the study group on database management systems. Inf. Syst. **3**(3), 173–191 (1978)
51. Webber, W.: Evaluating the effectiveness of keyword search. IEEE Data Eng. Bull. **33**(1), 55–60 (2010)
52. Weikum, G.: Where's the data in the big data wave? In: ACM SIGMOD Blog, March 2013. http://wp.sigmod.org/?p=786
53. Yu, J.X., Qin, L., Chang, L.: Keyword search in relational databases: a survey. IEEE Data Eng. Bull. **33**(1), 67–78 (2010)

Entity-Based Keyword Search
in Web Documents

Enrico Sartori[1], Yannis Velegrakis[1]([✉]), and Francesco Guerra[2]

[1] University of Trento, Trento, Italy
sartori.enrico@gmail.com, velgias@disi.unitn.eu
[2] University of Modena e Reggio Emilia, Modena, Italy
francesco.guerra@unimore.it

Abstract. In document search, documents are typically seen as a flat list of keywords. To deal with the syntactic interoperability, i.e., the use of different keywords to refer to the same real world entity, entity linkage has been used to replace keywords in the text with a unique identifier of the entity to which they are referring. Yet, the flat list of entities fails to capture the actual relationships that exist among the entities, information that is significant for a more effective document search. In this work we propose to go one step further from entity linkage in text, and model the documents as a set of structures that describe relationships among the entities mentioned in the text. We show that this kind of representation is significantly improving the effectiveness of document search. We describe the details of the implementation of the above idea and we present an extensive set of experimental results that prove our point.

1 Introduction

Most search engines for documents and news on the web are powered by a keyword-based indexing system. These indexing systems model the documents as a vector, i.e., a flat list of keywords. Keyword-based search is then achieved by looking at the relative frequencies of the various keywords in the documents and comparing them with the keywords in the user query, which is also modelled as a vector. Unfortunately, this approach is prone to the ambiguities met in natural languages, for instance, the use of different terms to describe the same real world entity. This is not simply due to the use of synonym terms, but extends to even keywords that are highly different. For instance, the term "Obama" and "US president" most likely refer to the same person, despite the fact that the two terms have no semantic similarity as words. Furthermore, it is also common the case in which the same keyword is used in different situations to describe highly different things. To cope with this problem, many systems employ entity linkage, i.e., the identification of the entity to which one or more consecutive keywords are referring and the replacement of these keywords with a unique reference to the entity. After this task a document can be seen as a vector of entities or a mixture of entities and keywords.

© Springer-Verlag Berlin Heidelberg 2016
N.T. Nguyen et al. (Eds.): TCCI XXI, LNCS 9630, pp. 21–49, 2016.
DOI: 10.1007/978-3-662-49521-6_2

Although the use of entities is significantly improving the accuracy of the search [21], there is still a major limitation: the loss of the relationships among the entities. Each entity in the vector representation of the document is seen independently of the others and its relationship is the same with any other entity in the vector. We advocate that by not seeing documents as a flat list of entities (and other keywords) but as a more structured whole, can lead to significantly better search results, and there is a need for a model that can represent and take this information into consideration.

In this work we provide a solution with a model that sees the document as a set of simple structures with no more than two entities each. Of course one can construct a complex structure like a graph or a tree [17] but then the complexity of the matching task is getting significantly higher. It is our belief that small structures specifying the relationships between entities found in a close proximity in the document alongside the kind of relationship that they have between them, provide the required additional information to achieve a satisfactory document search. Thus, the contributions of our work are as follows:

1. We introduce a model that instead of a vector of terms it uses a set of triples for representing documents and facilitating document search.
2. We illustrate how raw documents and user queries can be converted and modeled into that model.
3. We present an indexing mechanism that can effectively and efficiently index the structures we propose and as a consequence the documents themselves.
4. We propose a similarity metric that is used to identify the documents that are related to a given keyword query, and
5. We present a number of experimental results that demonstrate the efficiency and effectiveness of our proposed solution.

The remainder of this paper is structured as follows. Section 2 presents a motivating example that aims at helping the reader understand the importance of our approach and how it overcomes some of the limitations of the keyword search in documents. In Sect. 3 we define formally the problem we aim to address, while in Sect. 4 we explain in full details our solution. This includes our representation model, the indexing mechanism, the similarity formula, and the query answering algorithm. Section 5 describes the technical details of a prototype we have developed. Section 7 presents the related work and explains how we differ with what already exists. Finally, Sect. 6 contains the results of our experiments.

2   Motivating Example

Consider a user who is looking for a document talking about the position of the U.N. chief with respect to the situation in Syria and specifically its president Assad.

The document illustrated in Fig. 1 is an article that clearly talks about the topic that the user is looking for since the two main personalities, Ban Ki Moon who is the U.N. chief and Assad, the Syrian president, appear in the document.

Assad must stop violence, U.N. chief repeats

U.N. chief Ban Ki-moon on Monday renewed his appeal to Syrian President Bashar al-Assad to stop the violence against civilians after reports of a new military crackdown just a day earlier.

Ban, speaking to reporters in northeast Japan, said that he had delivered a strong message to Assad in a phone conversation on Saturday ahead of his report on Syria to the Security Council due on August 10.

"I hope he takes that situation very seriously and takes necessary measures respecting the will of the people," the U.N. Secretary General said when asked about latest signs that Assad continued to defy international pressure.

Residents of a Syrian provincial capital Deir al-Zor said government tanks stormed the city on Sunday, in the latest crackdown on five-month protests against President Bashar al-Assad's rule. Activists said 50 people had been killed in the attack, while the government denied the assault had taken place.

Syria has barred most journalists, making it hard to confirm events. The military assault on Deir al-Zor, about 400 km (250 miles) north-east of Damascus, was launched a day after Ban's conversation with the Syrian president.

Ban is visiting Japan's northeast nearly five months after a magnitude 9.0 earthquake unleashed a deadly tsunami that killed more than 20,000 and triggered the worst nuclear crisis since Chernobyl 25 years ago.

Fig. 1. An example document

Thus, the document would be expected to be among the results of the user query asking for the documents related to these two persons.

The challenging question to be answered, is how related the document is to the topic. Clearly the more a document mentions the topics of interest, in our case the two personalities, the more related it is. However, as it can be seen the two personalities are mentioned with different names, or with different expressions. For instance, Ban Ki Moon is mentioned some times with his name and some times as "U.N. Chief" or as "U.N. Secretary General". If the search is based only on the exact keywords that the user has used in the query, the relativeness of the document will depend on the frequency in the document of the keyword that was chosen to refer to a specific entity. To avoid these there is a need to identify and treat equally all the different expressions that refer to the same entity.

Furthermore, a different factor that significantly affects the relativeness of the document to the topic for which the user is looking, is the relationship between the two items of interest. In the example document, one can see how Ban Ki Moon and Assad appear often in the same sentence, this is a clear indication that in this article these two person are in a strong relationship. The last sentence, instead, references Japan, because the U.N. chief was there at the moment of his interview. Clearly, the relationship between Assad and Ban Ki Moon is much stronger than between Assad and Japan, even if they appear in the same document. Using a vector that treats all the keywords equally, assuming the same frequency, the strength between the two personalities of interest will be considered the same as the strength between the U.N. chief and the Japan. Thus, instead of only recognising entities in the document, it is important to understand the relationships that exist between them as mentioned in the document, and realise the strength of these relationships.

3   Problem Statement

Real world documents are sequences of keywords (or words for simplicity). To model a real world document we assume axiomatically the existence of a countable infinite set of *keywords* $\mathcal{W}$. Punctuation marks are also considered keywords.

Definition 1. *A* raw document D *is a finite sequence of keywords, i.e.,* $D = \langle k_1,$ $k_2, \ldots, k_n \rangle$*, where* $k_i \in \mathcal{W}$*, for* $i = 1..n$*, and* $n \in \mathbb{N}^*$*.*

From the semantic point of view, real world documents contain statements about real world entities. Statements describe characteristic attributes that entities have, actions the entities perform on other entities, or states in which they are. The actions or the states are typically expressed by verbs. To model these verbs, we consider a countable infinite set $\mathcal{V}$ of verb identifiers.

Entities, on the other hand, are referenced by noun phrases.[1] Since different noun phrases may refer to the same real world entity, it is often preferred to use a unique identifier for referencing the entity. Note that not all the noun phrases can be replaced by an entity identifier. For instance, for an expression "a red plane", or "an article" it is clear they talk about an entity but it is not clear what is the exact entity to which they are referring. To model this, we assume the existence of two countable infinite sets of identifiers, one $\mathcal{O}$ of entities, and one $\mathcal{N}$ for noun phrases. The set $\mathcal{O}$ contains one unique identifier for every real world entity. As a set $\mathcal{N}$ we consider the set of all the possible sequences of alphabet letters, digits and symbols. In this way, there is an easy way to find the identifier of a noun phrase. It only needs to create a string from the concatenation of the words in the noun phrase and use this concatenation as the noun identifier for the noun phrase.

We also consider the existence of a special identifier "*null*" that is the only identifier that at the same time is both an entity, a verb and a noun identifier, i.e., $null \in \mathcal{O}$, $null \in \mathcal{V}$, $null \in \mathcal{N}$ and $\mathcal{O} \cap \mathcal{V} \cap \mathcal{N} = \{null\}$. The *null* identifier functions differently than the way nulls function in databases. In particular any equality comparison of a null with another identifier or another null is always true. In other words, the null identifier functions like a wildcard.

Definition 2. *A* statement *is a triple* $\langle s, p, o \rangle$*, where* $s, o \in \mathcal{O} \cup \mathcal{N}$ *and* $p \in \mathcal{V}$*.*

We will use the symbol $\mathcal{S}$ to refer to the set of all possible statements.

Recent studies [8] on Wikipedia documents have indicated that considering only the entities mentioned in a document and the relationships between them preserves enough information to communicate with adequate accuracy the

---

[1] The meaning of a "noun phrase" is the one used in linguistics.

general semantic message of a document. Based on this, we also believe that entities (expressed through entity identifiers or noun phrase identifiers) and the relationships between them or their states (expressed through verb identifiers), provide adequate information to communicate the semantics of the content of a real world document to be successfully identified as related to the queries that ask for its information. For this reason, we define documents to be sets of statements.[2]

Definition 3. *A document is a finite set of statements, i.e., $D \subset S$ and is finite.*

We will use the symbol $\mathcal{D}$ to refer to the set of all possible documents.

Similar to real world documents, a user query is a finite list of keywords, only that its length, i.e., the number of keywords of which it consists, is significantly smaller than those of the documents. Thus, we can also consider the user queries as raw documents, and define the concept of a *query* as a document.

For simplicity, in what follows, and when there is no risk of confusion, we will drop the term "identifier" when talking about a verb, an entity, or a noun, respectively.

The problem we would like to solve is as follows. Given a collection of documents C^D and a user query q, provide an ordered list of documents in the collection, in a way that the first document is the one that is believed to be the most related to the query q, the second is the second most related, etc.

4 Solution

4.1 Processing the Documents

To convert the raw documents, i.e., the flat list of keywords of which the real world documents consist, into documents, i.e., set of statements, the very first step is to identify the keywords that refer to entities and replace them with the respective entity identifiers. This is achieved through a Named Entity Recognition process. Name Entity recognition is an extensively studied task and orthogonal to the scope of this work. Having identified the entities, the next step is to also identify the verbs. This is also orthogonal to our scope but can be achieved through syntactic and grammatic analysis, or a more complex Natural Language Processing in general, of the raw document text. After the verbs have been identified, they are replaced by the respective verb identifier for that verb. As verb identifier of a verb we consider its infinitive form. A syntactic and grammatic analysis is providing not only the verbs but also the noun phrases which are also replaced by their respective identifier. As a noun identifier for a noun phrase, it is considered the concatenation of the words that form the noun phrase, as already mentioned in the previous section. In this way the identification generation guarantees that two different noun phrases consisting of the same words

---

[2] Note that a "raw document" is what we defined as the document that the user provided, while a "document" is a set of statements containing identifiers and verbs.

Algorithm 1. Document Generation from Raw Text

Input: Raw Document D_r: $\langle k_1, k_2, .., k_n \rangle$

Output: Documemt D: Set of $\langle s, v, o \rangle$

DOCIMPORT(D_r)

```
(1)      (Set of Statements) D ← ∅
(2)      // Recognize entities and replace the keywords with the entity identifier
(3)      (Keyword Sequence) Dₑ ← NamedEntityRecognition(Dᵣ)
(4)      // Syntactic/Grammatical Analysis of the text and annotation of the various
         components.
(5)      (Keyword Sequence) Dₗ ← NaturalLanguageAnalysis(Dₑ)
(6)      (Set of Keyword Sequences) F ← SegmentIntoSentences(Dₗ)
(7)      foreach Dₛ∈F
(8)         for k=1 to |Dₛ|
(9)            v ← Dₛ[k]
(10)           if (v ∉ V)   continue
(11)           // Find the first verb on the left and the right of the verb v
(12)           lb ← 0
(13)           for i=1 to k-1
(14)              if (v ∈ V)   lb←i
(15)           la ← |Dₛ| + 1
(16)           for i=|Dₛ| to k+1
(17)              if (v ∈ V)   la←i
(18)           // Combine every id the left with every id from the right to form a
               statement
(19)           for i=lb+1 to k-1
(20)              for j=k+1 to la-1
(21)                 D ← D ∪ { ⟨Dₛ[i], v, Dₛ[j]⟩ }
(22)           // Special Cases
(23)           if |Dₛ|=1
(24)              D ← D ∪ { ⟨null, v, null⟩ }
(25)              continue
(26)           if (k=1 ∧ la−k>1)
(27)              for j=k+1 to la-1
(28)                 D ← D ∪ { ⟨null, v, Dₛ[j]⟩ }
(29)           if (k=|Dₛ| ∧ k−lb>1)
(30)              for j=lb+1 to k-1
(31)                 D ← D ∪ { ⟨Dₛ[j], v, null⟩ }
(32)        return (D)
```

but in different order, will be assigned a different noun identifiers. Any other keyword in the raw text that has not been identified as a verb, a noun or an entity, and is not a punctuation mark is eliminated.

The raw document has now been converted into a sequence of entity, verb, noun identifiers and punctuation marks. To convert the sequence into a set of statements, it is first segmented into sentences using the punctuation marks, which are then also eliminated. Every segment generated from the segmentation is a sequence of entity, verb and noun identifiers.

Since the verbs are the components that typically specify an action or a state, the verb identifiers are those that drive the statement generation process. In particular, for every verb identifier p in a segment we look at the entity or noun identifiers that appear on its left, i.e., before the verb identifier p in the sequence, and those on its right, i.e., those appearing in the sequence after the verb identifier. For each identifier s among those before, and every identifier o among those after, the statement $\langle s, v, o \rangle$ is created.

Since every segment corresponds to a sentence in the raw text, only one verb identifier will be typically present in every segment. However, since there are sentences in natural language with more than one verb, there are also segments with more than one verb identifiers. To cope with this case, when considering the noun and entity identifiers before the verb identifier p, we consider only those for which there is no other verb identifier between them and the p. Respectively, when considering the noun and entity identifiers after the verb identifier p, we consider only those for which there is no other verb identifier between them and the verb identifier p. As an example, consider the segment $\langle n_1, n_2, e_2, v_4, e_5, n_6, v_7, e_8, v_9, n_{10}, n_{11}, v_{12} \rangle$, where the v, e and n are meant to be verb, entity and noun identifiers, respectivelly. The statements that will be created containing the verb identifier v_7 with be the $\langle e_5, v_7, e_8 \rangle$ and the $\langle n_6, v_7, e_8 \rangle$, but not the $\langle e_2, v_7, e_8 \rangle$ because between the e_2 and the v_7 there is the v_6.

Of course there are special cases in which a statement as described above may not be created because the verb identifier is at the end or the beginning of the sequence or because there are two consequtive verbs. An example of such a situation is the v_{12} in the sequence above. For these cases, we use the special *null* identifier in the place of those missing. In the particular example of v_{12}, the statements that will be created are the $\langle n_{10}, v_{12}, null \rangle$ and the $\langle n_{11}, v_{12}, null \rangle$.

At the end of this step, the raw documents have become documents, i.e., the sequences of keywords have become sets of statements. The above steps of processing a raw document are illustrated in Algorithm 1.

4.2 Processing the User Query

Similar to the raw documents, the user queries have also to be brought into the document form. Recall that the user queries are flat lists of keywords, i.e., they are like raw documents. This means that a procedure similar to the one followed for the raw documents can also be followed here.

The first step that is performed is the identification of the verbs. Keyword queries typically have no verbs, but even if they have they are in some simple form. Thus, a simple lookup to a list of verbs is enough to identify if one or more keywords correspond to verbs, and replace them by the respective verb identifier.

The next step is the identification in the query of those keywords (or consecutive keywords) that refer to some real world entity and replace them with the respective entity identifier, if possible. This can be performed through a Named Entity Recognition task.

Algorithm 2. User Query to Set of Statements

Input: User Query Q_u: $\langle k_1, k_2, .., k_n \rangle$

Output: Query Q: Set of $\langle s, v, o \rangle$

QUERYPREP(Q_u)

(1)	(Query) $Q \leftarrow \emptyset$
(2)	(Sequence of keywords and identifiers) $D_e \leftarrow NamedEntityRecognition(Q_u)$
(3)	**for** i=1 **to** $|D_e|$
(4)	**if** $(D_e[i] \in \mathcal{O})$ **continue**
(5)	**if** $(lookupIfVerb(D_e[i]) \neq \emptyset)$
(6)	$D_e[i] \leftarrow$ generateVerbIdentifier($D_e[i]$)
(7)	**continue**;
(8)	**if** $(lookupIfNoun(D_e[i]) \neq \emptyset)$
(9)	$D_e[i] \leftarrow$ generateNounIdentifier($D_e[i]$)
(10)	// Replace each consequtive noun identifier set, with its powerset sequence
(11)	(Sequence of Identifiers) $T \leftarrow \emptyset$
(12)	(Sequence of Identifiers) $I \leftarrow \emptyset$
(13)	**for** i=1 **to** $|D_e|$
(14)	**if** $(D_e[i] \in \mathcal{O} \cup \mathcal{V})$
(15)	**if** $(T \neq \emptyset)$
(16)	(Set of Identifier Sequences) $P \leftarrow$ Powerset(T)
(17)	$T \leftarrow \emptyset$
(18)	$I \leftarrow I + P$
(19)	$I \leftarrow I + D_e[i]$
(20)	**else**
(21)	$T \leftarrow T + D_e[i]$
(22)	**for** k=1 **to** $|I|$
(23)	v $\leftarrow I[k]$
(24)	**if** $(v \notin \mathcal{V})$ **continue**
(25)	$lb \leftarrow 0$
(26)	**for** i=1 **to** k-1
(27)	**if** $(v \in \mathcal{V})$ $lb \leftarrow i$
(28)	$la \leftarrow |I| + 1$
(29)	**for** i=$|I|$ **to** k+1
(30)	**if** $(v \in$ V$)$ $la \leftarrow i$
(31)	**for** i=lb+1 **to** k-1
(32)	**for** j=k+1 **to** la-1
(33)	$Q \leftarrow Q \cup \{ \langle D_s[i], D_s[v], D_s[j] \rangle \}$
(34)	**if** $|I|$=1
(35)	**for** j=lb+1 **to** k-1
(36)	$Q \leftarrow Q \cup \{ \langle null, v, null \rangle \}$
(37)	**continue**
(38)	**if** $(k=1 \wedge la-k>1)$
(39)	**for** j=k+1 **to** la-1
(40)	$Q \leftarrow Q \cup \{ \langle null, v, D_s[j] \rangle \}$
(41)	**if** $k=|I| \wedge k-lb>1$
(42)	**for** j=lb+1 **to** k-1
(43)	$Q \leftarrow Q \cup \{ \langle D_s[j], v, null \rangle \}$
(44)	**return** (Q)

Next, we identify the noun phrases. Natural language analysis will not perform well in general here because of the brevity of the keyword queries and the lack of a complete syntax and grammar conformity. However, the only thing needed is to identify the noun words, which can be done without the full power of natural language processing. We can, for instance, simply do a lookup in a dictionary to identify if a word is a noun or not, or run the grammatical part only of a natural language processing. Apart from the verbs and the entity identifier that we already have, every non-noun word identified is ignored. Each noun words is then considered a noun phrase, and is replaced with its respective noun identifier.

Having converted the keyword query into a sequence of identifiers (verb, entity or noun), it is now possible to create a representation of it as a set of statements. The algorithm used for this purpose is the one used in documents as presented in the previous section. However, there is a small issue to be taken care, which introduces a small variation to the algorithm. Since it was not possible to run a full natural language analysis on the keyword query but we only characterized the keywords independently as nouns or not, there is the risk that two or more concequtive keywords that have been characterized as nouns, should have been considered not separately, but should form together a noun phrase. For instance, if the keyword query had the words "company restaurant", the partial natural language analysis that we can run would have only identify them as nouns. Then, according to the algorithm, each one should form a noun phrase, and be replaced by the respective noun identifier. However, it may be better the two nouns to be considered together as a single noun phrase. Unfortunately, we do not have enough information to decide which of the alternatives is the right one, so we consider them all during the statement generation phase. In the particular example, we would have considered three noun identifiers instead of two: the one for the "company", the one for "restaurant" and an additional one for the "company restaurant". Basically, if there are n concequtive keywords characterized as nouns, we consider all the $\frac{n*(n+1)}{2}$ ordered subsequences, and for each one we generate a noun identifier. Recall that the noun identifier for a sequence of work is the one generated by the concatenation of the words in the sequence. Each of the different noun identifiers we generate in this step is then used in the statement generation algorithm as if the respective noun phrase was present.

As an example, consider the user query:

$$q = \langle w_1,\ w_2,\ w_3,\ w_4,\ w_5,\ w_6 \rangle \tag{1}$$

which after the assignment of the identifiers becomes the sequence:

$$\langle n_1,\ n_2,\ v_3,\ n_4,\ v_5,\ n_6 \rangle \tag{2}$$

During the generation of the statements that involve the verb identifier v_3, the statements that will be generated will be the $\langle n_1, v_3, n_4 \rangle$, the $\langle n_2, v_3, n_4 \rangle$, and the $\langle n_{1-2}, v_3, v_4 \rangle$, where the noun identifier n_{1-2} is the one corresponding to the noun phrase consisting of both keywords w_1 and w_2.

Algorithm 3. Query Answering

Input: User Query Q_u: $\langle k_1, k_2, .., k_n \rangle$, Document Collection C^D
Output: Ordered List of Documents Ans
EVALUATE(Q_u)
(1)   (Sequence of $\langle Document, score \rangle$) $Ans \leftarrow \emptyset$
(2)   (Set of Documents) $L \leftarrow \emptyset$
(3)   (Query) $Q = QueryPrep(Q_u)$
(4)   **foreach** $D \in C^D$
(5)   int score $= ComputeScore(D, Q)$
(6) $Ans = Ans + \langle D, score \rangle$
(7) $Ans \leftarrow OrderByScore(Ans)$
(8)   **return** (Ans)

A special case that is important to mention here is the case in which the user query contains no verb, i.e., the phase that turns it into a sequence of identifiers produces a sequence of only entity and noun identifiers. In such a case we generate statements that have the special verb identifier "null". Since, in the absence of the verb, we are not sure what is the relationship between the entity and noun identifiers in the sequence, we generate one statement for every possible position among the keywords in the user query that a verb could have been present. For instance, if the user query was the

$$q = \langle w_1, \ w_2, \ w_3 \rangle \tag{3}$$

that after the identification assignment had become the sequence

$$q = \langle e_1, \ n_2, \ e_3 \rangle \tag{4}$$

the statements that would have been created are: $\langle null, null, e_1 \rangle$, the $\langle null, null, n_2 \rangle$, the $\langle null, null, e_3 \rangle$, the $\langle e_1, null, n_2 \rangle$, the $\langle e_1, null, e_3 \rangle$, the $\langle n_2, null, e_3 \rangle$, the $\langle e_1, null, null \rangle$, the $\langle n_2, null, null \rangle$, and the $\langle e_3, null, null \rangle$.

The steps described above are also explained in pseudocode in Algorithm 2.

4.3   Document and Query Matching

The main step of our work is to match the query to the documents and measure how related each document is to the query. Having both the documents and the query modeled as a set of statements, any kind of similarity across sets can be used. This would work well if every statement is treated as a monolithic and undivided object. However, this is not the case, or at least not the desired behavior for our approach. Imagine a user query that asks about Obama's visit to Middle East. Clearly a raw document that talks about such a visit is highly related. Although, a raw document that talks about Obama's visit to Russia, may not be what the user is looking for, it is not completely unrelated, since it is about Obama and also about a visit of him somewhere. Similarly, a raw document talking about some action Obama took in his government, is clearly less related, yet not completely irrelevant. The above mean that in addition to the set similarity metric that we need to employ, we have to also consider the partial matching statements. For this, given a document d and a statement t of the query q, we consider three sets of statements: The first set $\mathcal{A}$ consists of all those statements in the document that fully match some statement in the query q, i.e., the exact same statement appears in the query. The set $\mathcal{B}$ consists of all the statements in the document for which there is a statement in the query that matches two of their three components, i.e., either the subject and object, of the verb and one of the subject or object. The third set $\mathcal{C}$ consists of the document statements for which there is a statement in the query with which they match only one component, the subject, the verb, or the object.

Given these three sets, we compute the following three values:

$$s_1 = \frac{|\mathcal{A}|}{|S|}; \; s_2 = \frac{|\mathcal{B}|}{|S| - |\mathcal{A}|}; \; s_3 = \frac{|\mathcal{C}|}{|S| - (|\mathcal{A}| + |\mathcal{B}|)} \tag{5}$$

where the set S is the set of all the statements in the document and the lines around the set name indicate the cardinality of it, i.e., the number of statements it contains. The first number indicates the percentage of the document statements that fully match. The second number indicates the percentage of the partially matching statements with two components among those that are not matching fully. The third and last number indicate the percentage of the matching document statements of the document on one component with respect to the statements that are neither fully matched, nor partially matched with two elements.

With these three numbers we can compute a score of the matching of the statement t to the document d. The idea is that the percentage of the matching statements should count the most, those that are matching with only two components, less and those matching with only one component the least. Based on that principle, and a factor s which is between 0,5 and 1, we define the score of the document d to the statement t as:

$$s(t, d) = s_1 + (1 - s_1) \times [s_2 + (1 - s_2) \times s_3] \tag{6}$$

Many other formulas can be used for the computation of the $s(t, d)$, however, the above is the traditional weighted sum of two factors (with weights to give a sum to 1), but extended to capture three factors accordingly.

The final relatedness score of the document d to the quert q is the average of the matching scores of the query individual statements for that document, i.e.,

$$score(d, q) = \frac{\Sigma_{t \in q} s(t, d)}{|q|} \tag{7}$$

where $|q|$ denotes the number of statements in the query q.

The query evaluation steps all algorithmically explained in pseudocode in Algorithm 3.

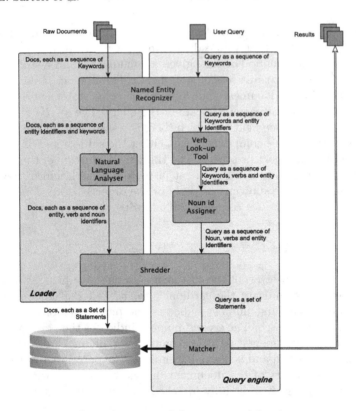

Fig. 2. The architecture of the query evalaluation system

5   Implementation

We have materialized the previously described framework into a system, the architecture of which is illustrated in Fig. 2. The figure illustrates clearly the different components and the group they belong depending on their role. In short, there is the part called Loader that is responsible for storing the documents into the document repository after turning each one of them from a sequence of keywords that is initially, into a set of statements. The figure illustrates the various components that are involved, which also reveals the flow of the process. The Loader is used offline, when new documents are to be added into the system, and in short, implements the task described in Algorithm 1. Another part of the system is the repository, illustrated as a disk, which is designed to store the documents in the form of sets of statements. A thirt part of the system is the Query Engine, which is the component working at run-time. Upon receiving a user query, it converts it into a set of statements by actually implementing Algorithm 2. Once the query has been converted to a set of statements, the Query Engine invokes the Matcher subcomponent that will compare this set to the set of statements of the documents in the repository and compute the matching scores for the stored documents, rank them, and return the ranked list of the matched documents to the user as an answer. These are the steps that Algorithm 3 describes. As with the case of the Loader, the flow illustrated in the architecture diagram for the Execution Engine reveals also the process flow.

In what follows we will provide a description of how each of these components of the system has been implemented and we will explain our design choices.

5.1   Named Entity Recognition

For the named entity recognition task we have opted for an external service instead of building a native solution. In particular, we are using the Open-Calais [27] which is freely available on the Internet. It is a mature, reliable and successful solution that can support our needs, without requiring an advanced pre-training that most of the other existing solutions require. Furthermore, the system is continuously updated and is guaranteed to perform in always new environments and communities.

A disadvantage of this solution is the latencies introduced by querying an external service over the network. To limit this disadvantage we adopted a "hybrid" implementation, i.e. to store in our database the phrases used to describe the entities encountered in the dataset. Intuitively, we have made our system to work as a cache of the expressions that have been so far met in the documents and mapped to entities by the OpenCalais. This has significantly reduced the latencies of this component. Furthermore this caching has a significant by-product advantage. Once it runs for the document collection, all the entities appearing in the document will be stored in the cache. When a new query arrives and the named entity identification runs on it, if for a sequence of keywords it is found that it is not present in the cache, it means that there is no document in the collection mentioning that entity, thus the query can be answered without any further delay.

OpenCalais is accessed through a REST interface using the method:

$$Enlighten(key, content, configuration)$$

which, given the HTML representation of a document, it cleans the text from the tags, and performs the Named Entity Recognition task, returning to the caller an XML representation of the result. The response from the service contains, for each entity, a *unique Id*, which is used by our system as well for the identification of the entity. Moreover, OpenCalais gives information about the position of the entity inside the analyzed text, and the sequence of keywords that describe it. This sequence of keywords is used in the cache entry that was previously mentioned.

Once the list of entities has been collected, the system marks the text of the document with the ids of the objects appearing into it. In practice it replaces the phrase describing the entity with the id, storing the information about the entity inside the database, to build the previously mentioned cache.

5.2   Natural Language Analysis

To identify the verbs and the nouns in the document sentences we need to perform some natural language text analysis. For this, we employed the NLTK [25], a well known and extensively used toolkit providing support for many common task in natural language processing.

As a first step, the document is tokenized by being divided into sentences and then into words. To perform the sentences isolation we relied on the algorithms provided by the NLTK toolkit, which allowed us to divide correctly any complex text. The tokenization algorithm takes a sentence and isolates the words appearing into it, preserving the position of the tokens inside the original sentence.

After this preprocessing step every sentence in the document is represented like a flat list of tokens, sorted according to their appearance in the sentence. Over this list is then applied a Part Of Speech tagging algorithm, which can recognize and annotate the different grammatical meaning of every token in the sentence. In our test implementation we relied again on the POS tagging algorithm provided by the NLTK. The main focus in this phase of the document processing is to tell apart nouns and verbs, in fact we need to recognize verbs, and link them with the keywords or the entities appearing in the noun part of the phrase. The output of this algorithm is a list of pairs (*token, POS tag*), once again maintaining the original order of the tokens in the analyzed sentence.

The POS tagging result is needed to construct a tree representing the structure of a sentence. In fact the list of POS tags is parsed and the system builds a tree of every sentence, matching the grammatical structure of the sentence. This tree is needed to resolve the link between verbs and nouns, trying to interpret the structure of the sentence, and connect the correct section of the phrase. For our purposes the aim is to obtain a tree which can be used to separate section of the phrase composed by nouns from the verbs and to follow the links between verbs and the nouns referring to it. To achieve this NLTK offers a parser, which by using as input a POS tags list can construct a tree of the sentence, querying a grammar given by the user. So we developed a grammar reflecting the structure of the information we need to extract from a natural language sentence. After this step, the sentence is in the form of a tree, representing the grammatical role of each token (noun, verb) and the relationship between them.

After the nouns and the verbs have been characterized in the text, the algorithms presented in the previous section for statement generation can be executed.

5.3   Repository Document Indexing

Since documents are sets of statements and a statement is a triple, we need a way to efficiently access the triples related to certain queries. To do so there is a need for an effective index structure. The index should help in the computation of the cardinality of the sets $\mathcal{A}$, $\mathcal{B}$ and $\mathcal{C}$ defined before by returning the number of triples that satisfy some search parameter. In particular, in order to compute the cardinality of set $\mathcal{A}$, the index should be able to compute the following function

$$i_{\mathcal{A}}(s, v, o, d) = cnt_{s,v,d,o} \tag{8}$$

where $cnt_{s,v,o,d}$ is the number of triples belonging to document d, that have the value s as *subject*, o as *object* and v as *verb*. For the cardinality of the set $\mathcal{B}$ it needs to be able to compute the function

$$i_{\mathcal{B}}(s, o, d) = cnt_{s,o,d} \tag{9}$$

where $cnt_{s,o,d}$ is the same as $cnt_{s,v,o,d}$, but with the v being of any value. Finally, the number of triples belonging to set $\mathcal{C}$, containing all the triples of the document that have a particular *subject* or *object*, requires the computation of the function:

$$i_{\mathcal{C}}(s, o, d) = cnt_{s,d} + cnt_{o,d} \tag{10}$$

where $cnt_{s,d}$ and $cnt_{o,d}$ count respectively the triples sharing only the *subject* with the query and only the *object*.

The index should be effective in terms of space in order to reside in main memory and achieve efficient look-ups. It should also support incremental updates when new documents are encountered by the system, without the need for a complete refresh of the whole structure through a from-scratch re-computation.

Basically our approach is derived from the Hexastore [28], we extended its algorithm and adapted it to fit our data and usage context. Hexastore is a system for the storage and querying of RDF triples. The idea behind it is to provide an indexing structure for every possible order of the three terms in a triple. It permits a fast filtering of the triples in a database, due to the ability to group the data with respect to any field in the triple.

The data structure we defined in order to implement a similar index is basically composed of a set of nested associative arrays, which can be navigated recursively using the fields of the triples as keys of the arrays. Figure 3 depicts how these nested arrays work. In this case, the structure permits to implement the function in formula 8. In fact, it is possible to follow the links from an array to the other and reach the counter associated with the last array. This field counts the number of triples in the document selected which have the *subject*, *object* and *verb* equal to the one searched. More specifically the first vector is an associative array whose keys are the set of different *subjects* encountered among all the triples. Each entry in this array is connected to a pair which is composed by a counter and another associative array. This structure is repeated recursively for the *object* field of the triple, and then for the verb. The cells of the last associative array are linked only to the respective counter.

Since our structure is fundamentally based on the Hexastore idea, the complexity of the retrieval of the respective tuples is the same to the one of Hexastore. A detailed study of its performance can be found elsewhere [28].

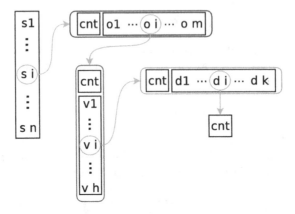

Fig. 3. Schematic representation of the index structure

To find the number of triples that match a triple over all their three fields in a document, we maintain and query a structure of the form:

$$[s_1 \ldots s_n] \to (c_s, [o_1 \ldots o_m]) \to$$
$$\to (c_{s,o}, [v_1 \ldots v_l]) \to$$
$$\to (c_{s,o,v}, [d_1 \ldots d_h]) \to$$
$$\to c_{s,o,v,d} \tag{11}$$

The notation above means that we have an array of the different subject values ($[s_1 \ldots s_n]$). For each entry in that array, i.e., for every subject s, we have a pointer to an array $[o_1 \ldots o_m]$, with each element of which corresponding to an object value from all those object values met in the triples that have s as a subject (c_s). An element of that array corresponding to an object value o is a pointer to an array $[v_1 \ldots v_l]$. Each element of that array corresponds to a verb v among all those verb values met in the triples that have subject s and object o. An element of the array $[v_1 \ldots v_l]$ corresponding to the verb v is a pointer to an array $[d_1 \ldots d_h]$, each element of which corresponds to a document in the collection, and indicates the number of triples of the form $\langle s, v, o \rangle$ that the respective document contains, i.e., the value of the function $c_{s,o,v,d}$. The latter is needed as described in formula (8) which in term is needed for the computation of the quantity $\mathcal{A}$ used in the formula (5).

In a similar fashion, to be able to quickly identify the number of triples in a document that match a specific subject and object, we maintain and use a structure of the form:

$$[s_1 \ldots s_n] \to (c_s, [o_1 \ldots o_m]) \to$$
$$\to (c_{s,o}, [d_1 \ldots d_h]) \to$$
$$\to c_{s,o,d} \tag{12}$$

which materializes the function $c_{s,o,d}$, used to compute the function mentioned in formula (9) which is required for the computation of the quantity $\mathcal{B}$ used in the formula (5).

Finally, for computing the function of the formula (10), we build and implement two structures, aimed at providing the numbers of triples in each document that match either the subject only or the object only. These structures are described in the formulas (13) and (14). They are used to retrieve the counts of triples matching at least one between *subject* and *object* fields of the given triple.

$$[s_1 \ldots s_n] \to (c_s, [d_1 \ldots d_m]) \to$$
$$\to (c_{s,d}) \tag{13}$$

$$[o_1 \ldots o_n] \to (c_o, [d_1 \ldots d_m]) \to$$
$$\to (c_{o,d}) \tag{14}$$

Note that the functions $c_{s,d}$ and $c_{o,d}$, include naturally the results of $c_{s,o,d}$ which in turn includes the results of $c_{s,o,v,d}$. The query answering algorithm, which we will present in the next subsection, will have to encounter for that fact.

The four structures that were just presented form our index structure. The index is built when the system starts, and is kept in memory throughout the operation of the system. Incremental updates are easily implemented due to the nesting of the structures and the use of associative arrays for the materialization of the respective array structures.

With the use of NLTK, the different keywords in the user query are tagged with the respective grammatical role, and the information about the position inside the raw document is maintained. The order of the words inside a query is important because we assume that, generally, when a user types a query, he or she tends to put verbs near to the nouns to which they refer.

5.4   Matcher

The matcher is invoked to identify the matching documents. In order to measure the relatedness of a document with respect to a specific statement in the user query, we need to identify the statement sets $\mathcal{A}$, $\mathcal{B}$ and $\mathcal{C}$ introduced in Sect. 4.3. The set $\mathcal{A}$ is built by a lookup in the structure represented in formula 11. Accessing the different associative arrays with the values of the triples fields, is possible to retrieve the list of documents containing a triple exactly like the one given. The second set is instead populated by the result of a visit in the structure in formula 12. In fact this structure returns the documents which have *subject* and *object* equal to the given triple. The counter associated with the document retrieved at this step will consider even the statements belonging to the set $\mathcal{A}$, therefore we need to subtract the previously computed result to obtain the right cardinality of the set $\mathcal{B}$. We are interested in knowing the cardinality of these sets, not in retrieving the specific statements, so we do not need to effectively isolate the statements of the set $\mathcal{B}$, but just keep track of the number of statements encountered. The computation of the cardinality of the set $\mathcal{C}$ is more complex and need two lookup in different structures. In fact here we need to count the statements that match only the *subject* or the *object* of the given statement. For the first subset the answer is provided by the structure in Eq. 13, while for the latter is used the formula in 14. The combination of these two subsets produce the set of statement that match the statement with *at least* one field among *subject* and *object*. In order to restrict the result only to the statements which match the statement with *only* one field it is necessary to subtract the cardinality of the previously computed sets. Now that the cardinality of each different class of similarity has been computed, it is possible to measure of relatedness of a document with respect to the statement given, using the equation in formula 6. The formula is computed for all the statements in the query, and then the results are averaged in order to compute the final measure of similarity between the query and the document. Once the results are computed, the list of documents is ranked with respect to the similarity score and returned to the user.

As previously mentioned, in our implementation, the index structures used to compute the cardinality of the three classes of similarity, return the values for all the documents containing the statement. In this way there is no need to compute the similarity for each document one by one, but for each statement is possible to retrieve the list of the documents related to that information. This approach has two important advantages: the first is that in this way we consider, and compute metrics, only for the documents related to the query, rather than analyzing one by one all the documents, considering even the ones totally unrelated to the query. The second is that by analyzing every single document independently, many statements will be compared multiple times, because they are shared among different documents.

6   Experiments

This section will describe the experimental methodologies and results obtained by testing our approach in terms of execution times, memory consumption and result quality. In Sect. 6.1 are presented the results obtained in terms of time performance in analyzing the data, and storing the information in the database. The next experiments, illustrated in Sect. 6.2, target the analysis of the time and memory requirements for the index structure, with respect to the dataset dimensions. Section 6.3, presents the performance of the query answering algorithm in terms of execution time. This experiment is divided into two parts, the first measures the response time with respect to the dimension of the query provided, while the second analyzes the answering time over the dimension of the dataset. Finally in Subsect. 6.4 we present the outcome of a comparison between a Lucene [18] search engine implementation (keyword-based) and our solution.

The dataset used to test the system is composed by 4000 documents, retrieved from newspaper websites. The documents have been collected regularly on a time window of some months, using the RSS feed of the newspaper to track the new documents published. The total size of the dataset is 321 Mb, which means that the average size of a document is 82.17 Kb, while, once the document is sent to OpenCalais, analyzed and stored into the database, its size shrinks to 23.56 Kbk and the whole dataset becomes something around 126 MB. We developed a web crawler able to read the RSS feeds and retrieve the documents published, keeping memory of the documents already retrieved (through a signature) in order to avoid duplicate data in the dataset. The documents were not selected from a particular domain an in general the OpenCalais was recognizing around 2–3 entities every 4–5 sentences. The nouns of the sentences that were not recognized as entities, as explained in the previous sections remained as is. The sentences had an average of 18 words in length. Notice that there is no correct translation of the documents (or the queries) into triples. As long as the matching task is sucessfully identifying the related documents.

The system tested in this section is an implementation of the approach written in Python 2. The advantages of this choice for the language are represented by the cross-platform feature of the language and the fast prototype developing time permitted by the richness of built-in structures featured by the language. The disadvantages are mainly the high resources consumption in terms of time and space that a Python implementation brings. The system was run on a computer running Ubuntu 14.04.2, with 8 GB of RAM and a 64bit, 2.40 GHz 1-core CPU.

6.1 Document Analysis

In this experiment is evaluated the time needed to analyze all the document in the dataset, and store the resulting data in the database. This corresponds to the first phase of the system lifecycle, and takes into consideration the time spent querying the OpenCalais service, and the analysis algorithm execution.

The plot in Fig. 4 shows the time taken by the system to perform the fetching and analysis of all the documents in the dataset over the dimension of the dataset itself. The higher of the two lines measure the time needed to fetch and analyze the documents, while the second is referred only to the analysis time, which depends entirely on our algorithm. The time needed to receive a response from OpenCalais, instead, depends on a variety of factors out of our control, mainly network latencies, document length and load condition on the remote service.

The execution time of our algorithm is itself divided in some subprocesses, which concur to this phase of the main process. In fact the system needs first to read the data from the machine filesystem, than it performs a preprocessing step. During this preprocessing step the text of the document is altered in order to mark the *entities* occurrences appearing in the article, and the relative information is prepared to be stored in the database. After this the effective grammatical analysis of the text is performed, which is the most time consuming operation in this phase. Once the algorithm terminates, the results are stored in the database.

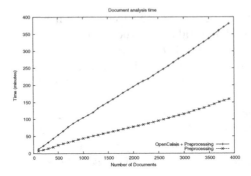

Fig. 4. Analysis time over dataset dimension

TASK	TIME
Query OpenCalais:	**3.375 s**
Read file:	0.001 s
Query service:	3.359 s
Storage time:	0.014 s
Analysis time:	**1.917 s**
Read file:	0.015 s
Entities marking:	0.004 s
Sentences analysis:	1.797 s
DB storage:	0.775 s

Fig. 5. Average Time of the individual tasks of processing a Document.

This plot shows how the execution time of the process is linearly dependent from the number of documents in the dataset, and in fact, from the experimental results evaluation, has been noticed that the time needed to analyze a single document is constant independently from the number of documents. The details of the time used to analyze a document are presented in table in Fig. 5, which shows how apart from the time spent waiting for the OpenCalais response, most of the time is spent in the analysis of the document's sentences.

6.2   Index Building

This experiment aims to show how the index, described in Sect. 5.3, performs in terms of time needed to fully populate the structures, and the memory consumption of the complete index. These features are evaluated with respect to the dimension of the dataset used, in order to underline the behavior of the system with growing amount of data. The plot in Fig. 6 shows how much time the system takes do build the index in relation with the dimension of the dataset.

The correlation between these two features is proportional to the number of documents multiplied by a small factor, because the time taken by the system to index a single document increments as the number of documents grows. This increment is given probably by the implementation of the Python dictionaries

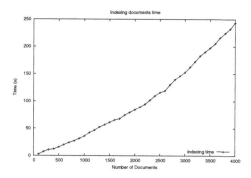

Fig. 6. Indexing time over dataset dimension

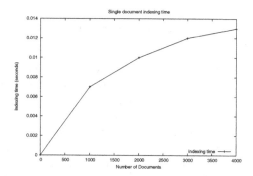

Fig. 7. Time taken to index a single document over dimension of the dataset

(associative arrays), used to implement the structure, which takes more time to complete the insertion operation as the dimension of the structure grows. The time needed to index a single document is reported in Fig. 7 in relation with the dataset dimension.

The other parameter which deserves to be considered in the evaluation of the index performances, is the footprint in main memory which maintaining this structure brings. As underlined in Sect. 5.3 the structure will contain all the data retrieved from the triples appearing across all the document in the dataset. For this reason the memory needed to store a similar structure is expected to grow linearly with the number of documents analyzed. In fact the plot in Fig. 8 shows this linear dependency between number of documents and memory used.

6.3   Query Answering

The aim of this experiment is to evaluate the performance of the query answering algorithm. In order to populate a set of queries which have at least a corresponding document in the dataset, we built a system which constructs automatically queries using information contained in the dataset. In particular we tested our

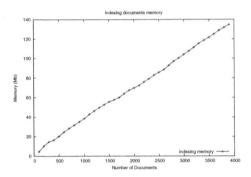

Fig. 8. Memory consumption of the index over number of documents

algorithm over three different kinds of queries, that we built following the approach described here:

- **Group 1:** The system selects a random document and retrieves the list of
  entities belonging to it. Form this list it extracts randomly a number (specified
  as a parameter) of entities. The query is than composed by concatenating a
  randomly chosen expression describing each entity. For our experiments we
  built queries with a number of entities ranging from 1 to 5.
- **Group 2:** To construct this set of queries a document is taken randomly, then
  a configurable number of triples is selected among the ones appearing in the
  document (for the experiment we used parameters ranging from 1 to 3). The
  content of the triples fields are than concatenated to form the query, and if one
  of the fields is an entity, it is replaced with an expression that corresponded to
  that entity (based on the information we have in the cache information from
  the OpenCalais).
- **Group 3:** It follows the same procedure as the previous group, with the
  difference that the entities are replaced not with any random expression that
  corresponds to them but to the one that actually appears in the selected
  document. For the last experiment as well we used a parameter for triples
  ranging from 1 to 3.

For each group of queries we run a series of experiments measuring the time
needed to build the query and the total response time used by the system to
provide the results. Moreover we kept track of how many triples are produced
from a keyword query. These results are presented by two graphs in Figs. 9 and
10, one showing the mean response and query building time, and the other
showing the distribution of the queries in terms of number of triples generated.
The time graph shows the mean response time over the number of triple, grouped
in buckets of width 50, while in the other the width of the buckets is 25.

From the graph in Fig. 9 is easy to evince that the response time grows
proportionally with the number of triples produced by the query, as it is expected
to behave. Moreover, the system can answer queries with up to 900 triples in less
than 0.5 s. The plot in Fig. 10, instead, shows how queries of different dimension

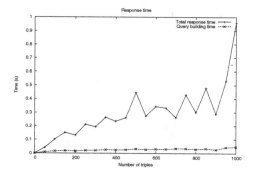

Fig. 9. Mean response time over number of triples in the query

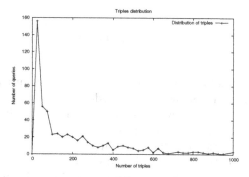

Fig. 10. Number of queries over number of triples in the query

(in terms of triples produced) are distributed. Is important to notice that most of the queries produce a number of triples less than 600 and after this value we encounter few queries. This helps to explain the behavior of graph Fig. 9. In fact, as the number of triples grows, the queries are fewer, and this brings to less predictable results in terms of mean response time.

6.4   Query Answering Quality

This section presents the outcome of the experiments run in order to measure the quality of the query answering algorithm. We implemented a test keyword-based search engine, and measured its performance against our approach, running the same queries on the two systems. The queries used are presented in Fig. 11(b), and have been selected by hand in an effort to simulate a human user behavior. The Fig. 11(a) presents the results of the quality comparison, showing in the first column the number of documents retrieved only by the keyword-based (Lucene) implementation, in the second column there is the number of documents retrieved by both systems and the third column gives the number of documents found only by our approach.

LUCENE	BOTH	ENTITY-BASED	QUERY ID
5 (0.6 %)	675 (92 %)	47 (6 %)	1
8 (0.8 %)	926 (96 %)	30 (3 %)	2
9 (2 %)	312 (90 %)	29 (8 %)	3
4 (0.9 %)	396 (94 %)	20 (4 %)	4
34 (10 %)	267 (79 %)	38 (11 %)	5

(a)

QUERY ID	KEYWORD QUERY
1	Saudi pressure Yemen
2	Sarkozy Ghaddafi Libya
3	English Premier League
4	Fukushima reactor meltdown
5	Bhopal gas leak

(b)

Fig. 11. Query answering quality evaluation results

From the queries results is possible to see that often an entity-based approach can retrieve more documents than a keyword-based one. The main reason for this is that once a group of keyword is resolved into an entity, this permits to collect the document that refers to the same entity using different keywords. An example of this situation can be the keyword *"Sarkozy"*, in the query no. 2, once resolved into an entity it can collect document which refers to *"president of France"*. These documents, instead are not retrieved by the keyword-based approach, because they do not contain the exact word *"Sarkozy"*.

A different situation happens in the fifth query, which aims to retrieve documents talking about the Bhopal disaster. The word "gas", appearing in the query is very common and permits to the keyword-based approach to find many other documents, even if they are less correlated to the query. The entity based one, instead, tries to resolve the set of keywords into the entity corresponding to the event, missing some documents which appear less correlated to the query.

The results show that few documents are retrieved only by the keyword-based engine even when our approach performs better. This behavior is explained by the different way in which our approach treats verbs and entities. While in the traditional approach verbs are just keywords, and the documents containing them are retrieved, in our approach verbs are considered only in relation to the entities they refer. Documents are retrieved if they contain the entities recognized in the query, while the presence of the exact verbs is used to enhance the result of a document. The only appearance of the verb in the document, without reference to the entities is not sufficient to consider the document relevant.

7   Related Work

What we present in this work constitutes information retrieval, since its main objective is to provide an approach to document retrieval and ranking approaches. The first approaches appearing in this field aimed to predict the probability distribution of terms inside documents, and hence determine the probability of relevance of a document with respect to a query. A very successful IR modeling approach is to convert the document into a vector space based representation, and therefore apply common techniques from linear algebra for computing similarities and operations on the documents. Generally this vector space is built using each word appearing across the dataset as a dimension of the space. A document in this space is represented as a vector, having as value for each dimension the frequency with which the corresponding word appears in the document. Once the documents are represented as vectors is possible to compute a measure of similarity between two of them using the *cosine similarity*. A broadly adopted document representation model, relying on the vector space, is the TF-IDF approach, which weights each word appearing across the dataset in a inversely proportional way with respect to the frequency of appearance of the word. This gives more importance to rare words appearing in a document, which will bring more relevance to the document if they match the user query. Roelleke and Wang [24] give a review of many aspects related to this approach to term relevance computation. An interesting addition to this approach is LSI (Latent semantic indexing), which permits to select the most informative words inside a document, making computation easier and removing noise. Moreover, this approach can isolate words that characterize a set of documents, enabling a semantic analysis of the content of these documents. An analysis of the performances of LSI, and some improvements at the existing approach are presented by Ando and Lee [3]. Since our concept is to identify the important words (entities, verbs and nouns) and not take into consideration other parts of the document apart from those three, in some way we are similar to LSI, that prioritise the words. Recently the ways of communication on the web has seen a broad evolution, with the average user becoming an active producer of information. To adapt to new situations and exploit additional data made available by websites, novel approaches to information retrieval have been presented. The model of publishing content used for blog, in which a single page can contain several documents, opens new possibility for finding different ways to deal with the problem to find the most suitable model to represent the documents. Moreover this kind of situation can shift the problem of document searching to blog searching based on the content of its posts. Another environment which offers different possibilities to model the information collected from webpages is Wikipedia, that enriches its content with metadata and links to other resources, this topics are investigated in [4].

Another important topic interested by this work regard the techniques used for named entity and entity management. One of the problems related to information retrieval, is that often the same entity has a number of different ways in which it can be referenced, this motivates the need of techniques that can identify uniquely objects on the web. An important contribution to this topic is the OKKAM project, which aims to integrate existing information about entities on the web and provide unique identifiers for them. Bouquet et al. [7] propose the use of an Entity Name System (ENS) with the aim to enable information collections to use unique identifiers for objects on the web. The use of entities can bring advantages in addressing different problems, as in data integration, where linking entities coming from different datasources is a major challenge. Ioannou et al. [15] present a query answering algorithm that permit to resolve entities and link together the information referring to them coming from different datasources.

As underlined in the rest of the paper we adapted a preexisting indexing algorithm to suit our needs for a fast way to access the data. The original work by Weiss et al. [28] addresses in particular the problem of indexing RDF data. In fact they consider this format as the main approach to represent semantic data retrieved from the Web. Their work aims to overcome the difficulties in terms of efficiency and scalability of managing large RDF databases, through the implementation of an indexing structure which allows fast query processing. The index is a set of nested associative arrays, which permit to group together, and therefore search, RDF triples with respect to their values. More conventional approaches to triple storage [11] and querying attempts to store the data in relational-like structures, but complex queries involve gathering information from multiple tables, worsening the performances of the system.

Sentence summarization techniques, used in this work to identify relationships between entities inside natural language, can be performed following different approaches. The goal of this task is to provide a context graph of the sentence, a formalism used to identify the semantic role played by the words into a natural language sentence. In this field, the development of ontologies of words available over the Internet enabled the application of approaches which relies on these platforms to perform their task. Hensman [13] present an algorithm that uses the WordNet database to construct context graphs from sentences. The approach followed is to retrieve from the ontology a set of possible phrase structure that the verb encountered can support and matching them with the sentence analyzed. Lei Zhang and Yong Yu [29], proposed a machine learning approach to address this problem, their work is focused on the context graph construction for sentence coming from a specific domain, in order to overcome the difficulties of dealing with general natural language sentences.

Understanding the semantics of the keyword query is a field that has received considerable attention [1,2,19,26]. Many works have studied the way to map the query keywords based on the database structures [14], or following some semantic approach without access to the database instance [5,6]. Many of these techniques can be used to enhance our approach, however, it is important to keep in mind that they do not actually employ any named entity identification techniques on the user query but they try to express the semantics of the keywords in terms of database structures. In that sense, they cannot be directly applied without some significant adaptation.

On the other hand, entity recognition in query has been sucessfully used before [12] and this task has been the driving force for the semantic search engines [10]. Named entities have not been used only for document retrieval but also other related data management tasks like indexing [20] and clustering [9] or matching in general [16]. For this reason, we believe that our querying technique is a significant complement to these works.

Finally, our idea of seeing the documents in a more structured form than a simple list of keywords has been studied in the past [17], however, the specific approach uses graphs as the document representation. We believe that graphs, despite very expressive, bring a unnecessary complexity than the sets of triples that we employ. In general, seeing the documents in a more structured way helps in better semantic search and in better matching what the user had in mind when formulating the query. As such, our approach can form an important foundation for further advancement of other query answering techniques like querying by example [22] or interactive query relaxation [23].

8 Conclusion

In this work we presented a different approach to document representation and query answering. The main features of the techniques we developed are the representation of a document in term of relationships individuated inside the text, and the resolution of groups of keywords in uniquely identified entity. This permits on one side to deduplicate the data, through the entity resolution, and the relationships based representation can increase the expressive power of the document, maintaining information about the actions performed by the entities appearing in the article. The experiments conducted on our testing implementation of this approach show how it scales in terms of time and memory consumption, and they prove the possibility to adopt a similar system to manage large datasets of documents. One of the possible environments of application of this work, as pointed out across the rest of the paper, is in fact the possible use inside a search engine for web documents, where the requirement of keeping a low response time with large datasets has a relevant importance.

References

1. Aditya, B., Bhalotia, G., Chakrabarti, S., Hulgeri, A., Nakhe, C., Parag, S.: Banks: browsing and keyword searching in relational databases. In: VLDB 2002, Proceedings of 28th International Conference on Very Large Data Bases, Hong Kong, China, 20-23 August 2002, pp. 1083–1086 (2002)
2. Agrawal, S., Chaudhuri, S., Das, G.: Dbxplorer: A system for keyword-based search over relational databases. In: Proceedings of the 18th International Conference on Data Engineering, San Jose, CA, USA, 26 February - 1 March 2002, pp. 5–16 (2002)
3. Ando, R.K., Lee, L.: Iterative residual rescaling: an analysis and generalization of lsi. In: Proceedings of the 24st Annual International ACM SIGIR Conference on Research and Development in Information Retrieval (2001)
4. Arguello, J., Elsas, J.L., Callan, J., Carbonell, J.G.: Document representation and query expansion models for blog recommendation. In: Association for the Advancement of Artificial Intelligence Conference (2008)

5. Bergamaschi, S., Guerra, F., Interlandi, M., Trillo-Lado, R., Velegrakis, Y.: Combining user and database perspective for solving keyword queries over relational databases. Inf. Syst. **2016**(55), 1–19 (2016)

6. Bergamaschi, S., Domnori, E., Guerra, F., Trillo-Lado, R., Velegrakis, Y.: Keyword search over relational databases: a metadata approach. In: Proceedings of the ACM SIGMOD International Conference on Management of Data, SIGMOD 2011, Athens, Greece, 12-16 June 2011, pp. 565–576 (2011)

7. Bouquet, P., Stoermer, H., Niederee, C., Mana, A.: Entity name system: The backbone of an open and scalable web of data. In: Proceedings of the IEEE International Conference on Semantic Computing, pp. 554–561 (2008)

8. Bykau, S., Korn, F., Srivastava, D., Velegrakis, Y.: Fine-grained controversy detection in wikipedia. In: 31st IEEE International Conference on Data Engineering, ICDE 2015, Seoul, South Korea, 13-17 April 2015, pp. 1573–1584 (2015). http://dx.doi.org/10.1109/ICDE.2015.7113426

9. Cao, T.H., Tang, T.M., Chau, C.K.: Text clustering with named entities: a model, experimentation and realization. In: Holmes, D.E., Jain, L.C. (eds.) Data Mining: Foundations and Intelligent Paradigms. ISRL, vol. 23, pp. 267–287. Springer, Heidelberg (2012)

10. Caputo, A., Basile, P., Semerato, G.: Integrating named entities in a semantic search engine. In: Proceedings of the 1st Italian Information Retrieval Workshop (2010)

11. Carroll, J.J., Dickinson, I., Dollin, C., Reynolds, D., Seaborne, A., Wilkinson, K.: Jena: implementing the semantic web recommendations. In: Proceedings of the 13th International World Wide Web Conference on Alternate Track Papers & Posters (2004)

12. Guo, J., Xu, G., Cheng, X., Li, H.: Named entity recognition in query. In: Proceedings of the 32nd Annual International ACM SIGIR Conference on Research and Development in Information Retrieval, SIGIR 2009, Boston, MA, USA, 19-23 July 2009, pp. 267–274 (2009). http://doi.acm.org/10.1145/1571941.1571989

13. Hensman, S.: Construction of conceptual graph representation of texts. In: HLT-SRWS 2004 Proceedings of the Student Research Workshop at HLT-NAACL (2004)

14. Hristidis, V., Papakonstantinou, Y.: Discover: keyword search in relational databases. In: VLDB 2002, Proceedings of 28th International Conference on Very Large Data Bases, Hong Kong, China, 20-23 August 2002, pp. 670–681 (2002)

15. Ioannou, E., Nejdl, W., Niedere, C., Velegrakis, Y.: On-the-fly entity-aware query processing in the presence of linkage. Proc. VLDB Endowment **3**, 429–438 (2010)

16. Ioannou, E., Rassadko, N., Velegrakis, Y.: On generating benchmark data for entity matching. J. Data Semant. **2**(1), 37–56 (2013)

17. Leskovec, J., Grobelnik, M., Milic-Frayling, N.: Learning sub-structures of document semantic graphs for document summarization. In: Workshop on Link Analysis and Group Detection (LinkKDD) (2004)

18. Lucene: https://lucene.apache.org

19. Luo, Y., Lin, X., Wang, W., Zhou, X.: Spark: top-k keyword query in relational databases. In: Proceedings of the ACM SIGMOD International Conference on Management of Data, Beijing, China, 12-14 June 2007, pp. 115–126. ACM (2007)

20. Mihalcea, R., Moldovan, D.: Document indexing using named entities (2001)

21. Mihalcea, R., Moldovan, D.I.: Document indexing using named entities. In: Studies in Informatics and Control (2001)

22. Mottin, D., Lissandrini, M., Velegrakis, Y., Palpanas, T.: Exemplar queries: Give me an example of what you need. PVLDB **7**(5), 365–376 (2014)

23. Mottin, D., Marascu, A., Roy, S.B., Das, G., Palpanas, T., Velegrakis, Y.: A probabilistic optimization framework for the empty-answer problem. PVLDB **6**(14), 1762–1773 (2013)
24. Roelleke, T., Wang, J.: Tf-idf uncovered: a study of theories and probabilities. In: Proceedings of the 31st Annual International ACM SIGIR conference on Research and Development in Information Retrieval (2008)
25. Steven, B., Loper, E., Klein, E.: Natural Language Processing with Python. O'Reilly Media Inc., Sebastopol (2009)
26. Tata, S., Lohman, G.M.: SQAK: doing more with keywords. In: Proceedings of the ACM SIGMOD International Conference on Management of Data, SIGMOD 2008, Vancouver, BC, Canada, 10-12 June 2008, pp. 889–902. ACM (2008)
27. The OpenCalais System: http://www.opencalais.com
28. Weiss, C., Karras, P., Bernstein, A.: Hexastore: sextuple indexing for semantic web data management. Proc. VLDB Endowment **1**, 1008–1019 (2008)
29. Zhang, L., Yu, Y.: Learning to generate CGs from domain specific sentences. In: Delugach, H.S., Stumme, G. (eds.) ICCS 2001. LNCS (LNAI), vol. 2120, pp. 44–57. Springer, Heidelberg (2001)

Evaluation of Keyword Search in Affective Multimedia Databases

Marko Horvat[1]([✉]), Marin Vuković[2], and Željka Car[2]

[1] Department of Computer Science and Information Technology,
Zagreb University of Applied Sciences, Vrbik 8, 10000 Zagreb, Croatia
`Marko.Horvat@tvz.hr`
[2] Faculty of Electrical Engineering and Computing,
University of Zagreb, Unska 3, 10000 Zagreb, Croatia
`{Marin.Vukovic,Zeljka.Car}@fer.hr`

Abstract. Multimedia documents such as pictures, videos, sounds and text provoke emotional responses of different intensity and polarity. These stimuli are stored in affective multimedia databases together with description of their semantics based on keywords from unsupervised glossaries, expected emotion elicitation potential and other important contextual information. Affective multimedia databases are important in many different areas of research, such as affective computing, human-computer interaction and cognitive sciences, where it is necessary to deliberately modulate emotional states of individuals. However, restrictions in the employed semantic data models impair retrieval performance measures thus severely limiting the databases' overall usability. An experimental evaluation of multi-keyword search in affective multimedia databases, using lift charts as binomial classifiers optimized for retrieval precision or sensitivity, is presented. Suggestions for improving expressiveness and formality of data models are elaborated, as well as introduction of dedicated ontologies which could lead to better data interoperability.

Keywords: Affective multimedia · Information retrieval · Classification · Semantic annotation · Emotion · Lexical similarity

1 Introduction

Affective multimedia databases are repositories of multimedia documents with annotated semantic and emotion information. Although important and having many practical applications the design of these databases is not standardized and varies greatly. In general they all consist of a digital storage and a formatted document containing metadata about documents' high-level semantics and expected emotional states that will be provoked in a human subject when exposed to a particular multimedia document stored in the database. Generally, affective multimedia databases can be searched with text-based and content-based paradigms. The text-based search may use free-text keywords

© Springer-Verlag Berlin Heidelberg 2016
N.T. Nguyen et al. (Eds.): TCCI XXI, LNCS 9630, pp. 50–68, 2016.
DOI: 10.1007/978-3-662-49521-6_3

describing multimedia semantics and affective annotations in the form of numeric values or free and structured text. Content-based search relies on the multimedia documents themselves and their digital content from which it is possible to extract information such as colorization, edges in the image, visual regions, blobs, shapes, and finally objects and their mutual spatial relationships.

Affective multimedia databases are a specific subgroup of general-purpose multimedia repositories with several important distinctive characteristics, elaborated in the next section, that lead to inadequate use of semantic retrieval. These databases use sparse annotation models with limited and disjunctive sets of unrestricted keywords. A multimedia document is often described with a single free-text keyword and only rarely with more than one keyword. Furthermore, a large proportion of keywords is unique for a specific document and do not describe other documents in the database. In such a model it is necessary to semantically query the database with a narrowly defined set of keywords [1]. Indeed, to improve retrieval efficiency – from the user's standpoint – it is beneficial to become an expert in specific databases and come to know the employed keywords and specifics of each database model. As a consequence some practitioners avoid semantically querying affective multimedia databases altogether and use other methods to retrieve the desired content, although these may require additional effort. As has been shown in a survey on usage modalities of emotionally-annotated picture databases [2] in practice it is more productive to browse affective multimedia databases visually by examining the digital storage content, although they may contain hundreds or thousands of images, than to perform keyword search. Only 23.33 % of researches retrieve stimuli both manually and with a software tool. Most participants (60 %) deem stimuli descriptors to be at least in some way inadequate, ambiguous or insufficient in conveying the true semantic and emotional multimedia content. Additionally, the survey showed that 70 % of participating experts need more than 1 h to retrieve optimal pictures for a single emotion elicitation sequence and 20 % require more than 12 h which is extremely impractical. The survey indicates that manual search for the most appropriate visual stimuli in a database with over thousand different images – each with its own specific visual properties, semantic and emotional context – is not trivial.

Choosing documents for optimal stimulation of emotional reactions is a time-consuming and labor-intensive activity. In order to alleviate this task we believe that special attention should be given to introduction of efficient keyword search methods to affective multimedia systems. This popular and well-researched retrieval modality is currently largely neglected in affective multimedia databases. The contribution of the paper lies in providing a evaluation of keyword search effectiveness in a typical and commonly used affective multimedia database. Results of such evaluation provide helpful directions for development of software tools for querying and extraction of multimedia stimuli.

The remainder of this paper is organized as follows; Sect. 2 gives a detailed overview of the most important contemporary affective multimedia databases with their semantic and emotion models. Important differences between affective multimedia databases and other multimedia databases are elaborated in this section. Keyword search specificities in affective multimedia databases are also explained in Sect. 2. The employed emotion theories and models for annotation of multimedia semantics are illustrated – with examples – in Sects. 2.1 and 2.2, respectively. Section 3 describes the undertaken evaluation of keyword search in affective multimedia databases with an overview of the experimental dataset in Sect. 3.1. Relatedness measures between search queries and picture metadata annotations, ranking algorithms, as well as the lift chart binomial classifier are described in Sect. 3.2. The obtained results are presented in detail and discussed in Sect. 3.3. The related work is brought forward in Sect. 4. Finally, Sect. 5 concludes the paper and provides an outlook to future activities regarding the keyword search in affective multimedia databases.

2   Affective Multimedia Databases

Any multimedia file can generate positive, negative or neutral emotions of varying intensity and duration [3]. Multimedia documents with a priori annotated semantic and emotion content are stored in affective multimedia databases and are intended for inducing or stimulating emotions in exposed subjects. Because of their purpose such affective multimedia documents are also referred to as stimuli. By observing still images, films, printed text or listening to sounds, music and voices emotional states of affected subjects may be modulated [4, 5]. This spontaneous cognitive process is an important research topic in psychology, neuroscience and cognitive sciences but also in many interdisciplinary domains like Affective Computing and Human-Computer Interaction (HCI). Affective multimedia databases are particularly useful in research of emotion and attention, but they can also be employed in the study of stress-related mental disorders [6]. To fully understand particularities and limitations of affective multimedia databases, and why keyword search – as analyzed in Sect. 3 – is not more prevalent, it is first necessary to consider the their architecture and content.

The affective multimedia databases can be distinguished from other multimedia repositories foremost by the purpose of their content (i.e. to purposely elicit specific emotional reactions), but also by the models employed to describe knowledge about emotions and by rigidly controlled experiments with which emotion metadata is acquired. The emotion models (Sect. 2.1) are standardized which allows stimuli to be used in a controllable and predictable manner: the emotion elicitation results can be measured, replicated and validated by different research teams [7, 8]. On the other hand, semantics of affective documents (Sect. 2.2) are usually described very rudimentary, especially compared to current standards in commonplace internet multimedia repositories or social webs. Affective multimedia databases represent useful tools to practitioners because combined with immersive and unobtrusive visualization hardware in low-interference ambient they provide a simple, low-cost and efficient means to investigate a wide range of emotional reactions [9, 10].

Currently many affective multimedia databases exists but International Affective Picture System (IAPS) [7] and International Affective Digital Sounds system (IADS) [8] are two of the most cited databases in the area of affective stimulation. They cover a wide range of semantic categories characterized along the affective dimensions of pleasure, arousal and dominance. At the moment of writing Nencki Affective Picture System (NAPS) is the largest and the most recently developed database [11].

Fig. 1. Representative picture stimuli from often used the IAPS, GAPED, NAPS and NimStim affective multimedia databases. Emotion and semantics are displayed below each stimulus.

It features significant models improvements compared such as semantic categories, multi-word stimuli annotations and also incorporated discrete (emotion norms) and dimensional emotion models. The IAPS, IADS and NAPS contain 1182, 167 and 1356 semantically and emotionally annotated stimuli, respectively. Some of the common visual stimuli are displayed in Fig. 1.

Emotion annotations in affective multimedia databases are always experimentally acquired and verified. For example during development of IAPS subjective ratings from around one hundred participants have been acquired for each picture and statistically analyzed to acquire aggregated mean and standard deviation of emotional dimension values [7]. Other multimedia databases follow the same pattern.

The databases were created to achieve three objectives [12]: (1) better experimental control of emotional stimuli, (2) increase the ability of cross-study comparisons of results; and (3) facilitate direct replication of undertaken studies. These principles are shared among other affective multimedia databases. Apart from the IAPS and IADS the most frequently used and readily available affective multimedia databases are Nencki Affective Pictures System (NAPS), Geneva Affective PicturE Database (GAPED) [13], Dataset for Emotion Analysis using EEG, Physiological and video signals (DEAP) [14], NimStim Face Stimulus Set [15], Pictures of Facial Affect (POFA) [16], Affective Norms for English Words (ANEW) [17] and SentiWordNet [18]. Additional audio-visual affective multimedia databases with category or dimensional emotion annotations are listed here [19]. Pictures are the most represented multimedia format and facial expressions are the most numerous database type. Facial expression databases are primarily used for face recognition and face detection, but they can be also successfully employed in emotion elicitation with minimal context. A more detailed overview of facial expression databases is available in literature [20].

In summary, available standardized affective multimedia databases offer a variety of quality audio-visual stimuli to researchers in the field. They were meticulously built in controlled experiments and enable valid comparative measurements. But also they are mutually noncompliant, non-contiguous and structurally diversified. Their content, multimedia document semantics in particular, is described loosely and informally.

2.1 Models of Multimedia Affect

In computer systems two predominant theories are used to describe knowledge about affect: the discrete category model and the dimensional model [21]. The former model is also referred to as emotion norms or basic emotions, while the latter is also called Circumplex model of affect [22] or Pleasure Arousal Dominance (PAD) [23]. Both theories of affect can effectively describe emotion in digital systems but are not mutually exclusive. All affective multimedia databases have been characterized according to at least one of these models [21], and for some – such as the IAPS, IADS, ANEW and NAPS – data from both models are available. Annotations from both models are useful because they provide a more complete characterization of stimuli affect.

The dimensional theories of emotion propose that affective meaning can be well characterized by a small number of dimensions. Dimensions are chosen on their ability to statistically characterize subjective emotional ratings with the least number of dimensions possible [24]. The dimensional model is built around three emotion dimensions: valence (*Val*), arousal (*Ar*) and dominance (*Dom*). Frequently only first two are used because dominance is the least informative measure of the elicited affect [21]. All three dimensions are described with continuous variables $al \in [1, 9] \in Val, ar \in [1, 9] \in Ar,$ $dom \in [1, 9] \in Dom$. In some databases emotions values are scaled and represented in percentages. Therefore, for all practical purposes a single multimedia document can be represented as a coordinate in a two-dimensional emotion space $\Omega_{Emo} = Val \times Ar$ as in Fig. 2.

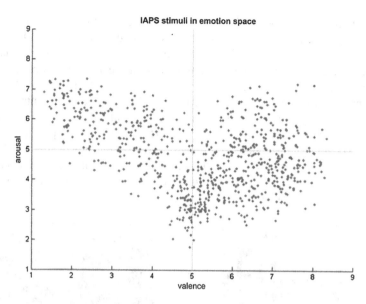

Fig. 2. Average dimensional model values of 1182 IAPS stimuli in valence-arousal emotion space.

Positivity and negativity are specified by valence, arousal describes the intensity or energy level of a stimulus while dominance represents the controlling and dominant nature of the emotion. In contrast to the dimensional theories, categorical theories claim that the dimensional models do not accurately reflect the neural systems underlying emotional responses. Instead, supporters of these theories propose that there are a number of emotions that are universal across cultures and have an evolutionary and biological basis [25].

Which discrete emotions are included in these theories is a point of contention, as is the choice of which dimensions to include in the dimensional models. Most supporters of discrete emotion theories agree that at least the five primary emotions of happiness, sadness, anger, fear and disgust should be included.

2.2 Keyword-Based Semantic Model in Affective Multimedia Databases

In almost all affective multimedia databases a multimedia stimulus is described with a single tag from an unsupervised glossary. Semantic relations between different concepts are undefined and multiple lexically different keywords are often used for description of synonym or similar concepts. For example, in the IAPS a picture stimulus portraying an attack dog is equally tagged as "dog", "attack", "attackdog" and "attack_dog". Without support for natural language processing and higher knowledge synonym keywords like "canine" or "hound" would be interpreted as different concepts just like singulars and plurals "man" and "men" or "woman" and "women". Incorporated semantic models do not implement semantic similarity measures and there are no criteria to estimate relatedness between concepts. In such databases it is impossible to establish that, for example, "dog" and "cat" are more closely related than "dog" and "orange". This represents significant drawback because a search query has to lexically match the database's keywords. Since currently semantic models in affective multimedia databases use free-text annotations only lexical interpretation of meaning of search queries and multimedia metadata is possible. More expressive and formal semantic models are not possible without modification of database multimedia descriptors and introduction of appropriate knowledge structures. The inadequate semantic descriptors result in three negative effects which impair stimuli retrieval: (1) low recall, (2) low precision and high recall or (3) vocabulary mismatch. Additionally, affective multimedia databases are not integrated with knowledge bases that describe high level multimedia semantics. An example of the inadequacy of semantic descriptors in contemporary affective multimedia databases is illustrated in the Fig. 3.

IAPS, 5635, "WinterStreet", valence=6.25, arousal=3.97   IAPS, 7039, "Train", valence=5.93, arousal= 3.29

Fig. 3. An example of two IAPS pictures (5635.jpg, 7039.jpg) with similar emotion annotations but different and inadequately described high-level semantics. Stimuli with complex contexts cannot be sufficiently well described with existing models.

Pictures p_1 and p_2 are stored in the IAPS database ($p_1 = 5635$.jpg, $p_2 = 7039$.jpg) and have dimensional emotion values. Both pictures have closely similar neutral valence and below neutral arousal $p_1: (val, ar) = (6.25, 3.97)$, $p_2: (val, ar) = (5.93, 3.29)$. Neutral values of emotional dimensions do not provoke positive and negative valence or arousal responses. Typically, neutral values range between 4.5–6.5 for both dimensions.

Although both pictures have similar valence and arousal they have different semantics and would be appraised differently by subjects. High-level semantics of p_1 and p_2 are described with just one keyword "WinterStreet" and "Train", respectively. While this may be enough for representation of an isolated object, e.g. in a close-up picture with minimal context, this is clearly insufficient for rich meaning of objects within a scene even just for the most relevant and evident high-level semantics.

For example, in IAPS picture p_1 is annotated with a single keyword "WinterStreet" but any of the following tags could substitute the original keyword: "SnowedIn", "WinterCity", "SnowCoveredCars" or even "Desolation". Equally so, picture p_2 could be described as "FreightTrain", "Railroad", "Scenery", "Autumn" or "Beauty". Reliance on emotion values alone is not enough to discriminate stimuli and IAPS semantic model is insufficient in complete description of multimedia content, and other affective multimedia database share the same annotation scheme.

3 Multimedia Keyword Search Experiment

In order to explore the practicality of keyword-based retrieval in existing affective multimedia databases it was necessary to conduct an experiment using typical semantic models and usage scenarios. Since emotions in affective multimedia databases are described as vectors in Euclidian space [22, 23] with well-known distribution while high-level multimedia meaning is more uncertain, semantics is a more important parameter in evaluation. The dataset was extracted from the IAPS [7] since its semantic model (as described in Sect. 2.2) is at the least equally, or more, sophisticated than that of other databases. Additionally, the IAPS is the largest, most popular and intensively used affective multimedia databases in the world. As such the IAPS represents the best source of experimental material because the evaluation will be pertinent to other affective multimedia databases and will have an important practical significance.

In the keyword search evaluation experiment a set of 741 pictures was first extracted from the IAPS and then repetitively queried with one, two and three unmanaged keywords from the set using two different but common lexical similarity metrics. Each time a subset of 100 pictures was randomly selected from the larger set. Naïve and approximate string matching algorithms were used to sort the results. The pictures were ranked according to similarity of their tags relative to keywords posited in the query and binary classified as true or false using lift charts. Information retrieval measures (accuracy, precision, recall, fall-out and F-measure) for each retrieval task were aggregated and analyzed. The details of this evaluation are described in Sects. 3.1 and 3.2 while the results are presented in Sect. 3.3.

The experiment was accomplished with a custom-made software tool Intelligent Stimuli Generator (intStimGen). The tool (Fig. 4) was designed as n-layered Desktop application with an efficient user-friendly graphical interface. The tool is written in .NET 3.5 framework and optionally may use Jena.NET toolkit for querying and integration with the Protégé ontology editor. The intStimGen is designed for keyword and ontology retrieval in affective multimedia databases. In the current version it supports integrated search of the IAPS and IADS databases using semantic and emotion descriptors, construction of visual and auditory stimuli sequences and presentation of sequence series to subjects. Elicited physiological and behavioral responses may be acquired using biometric sensors or cameras. For the purpose of this experiment only features related to text-based search were used.

With the intStimGen relatedness between semantic descriptors can be defined using lexical and semantic similarity measures. The implemented measures are: Levenshtein, path length, Wu Palmer, Leacock Chodorow and Li algorithm [26]. Other measures can be modularly added later if needed. Document retrieval and ranking can be performed using any combination of descriptors and similarity measures.

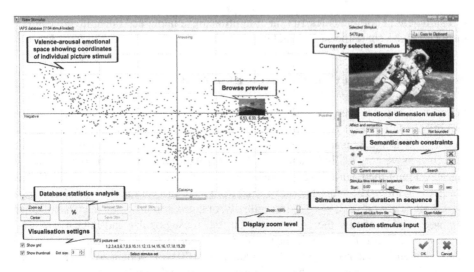

Fig. 4. The search screen of the Intelligent Stimuli Generator tool used for the evaluation of keyword search displaying the IAPS affective multimedia database.

3.1 Dataset

For experimentation in keyword multimedia retrieval a set of 741 pictures were extracted from the IAPS database, as outlined in the previous section. The selected pictures are mostly semantically unambiguous (i.e. they have a clear and easily comprehensible meaning) and do not provoke intense negative emotions but they do have different levels of arousal. Each picture is described with one keyword such as "Accident", "Adult", "AimedGun", "Aircraft", "Airplane", "Biking/train", "BingeEating", "Boy", "Boys", "BoysReading", "BoysW/Guns", "BurningCar", "BurntBldg" etc. As can be seen, some are acronyms from two or three words and even contain special characters, while some are proper words from the English language.

The selected pictures show animals, buildings, nature, people, sport activities, household objects and food. Each picture in document collection **D** was annotated with an original IAPS keyword $w_i \in$ **W**. In total, the selected pictures were described with 387 different keywords. The selected images are emotionally either neutral or highly positive with varying levels of arousal. Picture affective values were inherited from the IAPS. The set of all queries **Q** consisted of different individual queries exclusively from the annotation glossary **W** (**Q** $\subseteq$ **W**). For each single query $q_i \in$ **Q** a subset of documents $\dot{D} \subset$ **D** with $|\dot{D}| = 100$ pictures were randomly selected and classified using the afore-mentioned glossary and two relatedness measures. The queries contained one, two and three keywords. One term is minimally sufficient to produce a search and at least three different terms were included in metadata of all document subsets used in the evaluation. Since IAPS pictures predominantly have clear meaning it is not always possible to assign many keywords to their context. Only terms existing in the annotations of **D** were used as queries thereby ensuring that the retrieved set should be nonempty. In total 60 queries on sets of randomly selected pictures were executed and 6000 samples were classified twice for precision and sensitivity. Since the classification was binary, i.e. has only two classes — as described in the next section, the term sensitivity is used instead of recall.

The goal of classification for precision is to maximize the fraction of retrieved pictures that are relevant in $\dot{\mathbf{D}}$ for each $q_i \in \mathbf{Q}$, while classification for sensitivity will try the opposite, namely to maximize the fraction of relevant pictures that are retrieved in $\dot{\mathbf{D}}$, also for every $q_i \in \mathbf{Q}$. In practice the former optimization method will have high accuracy but will result in a small number of retrieved instances, and the latter optimization will return a larger fraction of instances but with considerably lower accuracy. Generally speaking, an ideal classification should do both: have high accuracy and high recall.

3.2   Lexical Matching, Ranking and Classification

The retrieved pictures were ranked using string inclusion and Levenshtein algorithms [27]. String inclusion is also called naïve string search algorithm, and Levenshtein edit distance metrics. Each of the ranking algorithms provided a measure of lexical relatedness (i.e. similarity score) between two labels a, b as rel (a, b) $\in$ [0, 1], where a, b $\in$ **W**, with properties

$$\text{rel}(a, b) = 1, \ x = y$$
$$\text{rel}(a, b) < 1, \ x \neq y \tag{1}$$

Since unsupervised keywords are mutually semantically unrelated only lexical similarity algorithms (such as string inclusion and Levenshtein) can be used to establish a relatedness measure between them. The inclusion represents the simplest measure that only checks if a string is included or exists in another string. Since this is an exact lexical matching measure its output is binary (0 or 1) either two strings do not match at all, or they do match completely. Naïve search and Levenshtein algorithms were chosen for the experiment as characteristic representatives of two different types of lexical searching algorithms – exact and approximate, respectively. The former do not allow errors in search terms unlike the latter. Naïve method is rudimentary and can be the most easily implemented in all existing affective multimedia databases. Levenshtein is a well-known approximate search algorithm representing a whole family of edit distance lexical metrics. It implementation is freely available in many programming languages, application frameworks and computer tools.

For example, in the experiment one picture 2341.jpg is tagged t = "Children" and 1-word search query was q_1 = "Child". The naïve method (i.e. exact matching) resulted in $\text{rel}_{exact}(t, q_1) = 0$ since t and q_1 were not completely identical. For the same example Levenshtein measure, as approximate lexical matching algorithm, was $\text{rel}_{approx}(t, q_1) = 0.625$. For a 2-word query $q_2 = (q_2^1 \lor q_2^2)$, where q_2^1 = "Child" and q_2^2 = "Children", and the same picture tag t, an aggregation of individual relatedness scores was necessary. In the experiment arithmetic mean was used as the aggregation function. Therefore, for the 2-word query q_2 the exact matching was

$$\text{rel}_{exact}\left(t, q_2\right) = 1/2 \left(\text{rel}_{exact}\left(t, q_2^1\right) + \text{rel}_{exact}\left(t, q_2^2\right)\right) = 1/2\left(0 + 1\right) = 0.5 \qquad (2)$$

while approximate metrics provided a higher score

$$\text{rel}_{approx}\left(t, q_2\right) = 1/2 \left(\text{rel}_{approx}\left(t, q_2^1\right) + \text{rel}_{approx}\left(t, q_2^2\right)\right) = 1/2\left(0.625 + 1\right) = 0.8125 \qquad (3)$$

Since each query q_i always returned all $|\dot{\mathbf{D}}| = 100$ images (there was no cut-off) in descending order, with picture most relevant to the search query on top of the list and the least relevant at the end, it was necessary to determine the threshold value t that determines the classification boundary and how the retrieved pictures will be categorized. Given a retrieved picture $p_i \in \dot{\mathbf{D}}$ and its rank in the set of retrieved pictures $r_i \in [1, 100]$, where $r_j = 1$ indicates the most relevant picture p_j, the picture p_i was classified in category $cat_i = \{True, False\}$ as

$$cat_i = \text{True}, \ r_i \leq t$$
$$cat_i = \text{False}, \ r_i > t \qquad (4)$$

Therefore, in each query the retrieved corpus was binary classified in two subsets: pictures relevant (category True) and irrelevant (category False) to the search. The threshold value was determined for each query using a lift chart. For the maximum precision in retrieval the threshold was set to the rank with the highest lift factor. This adaptive approach enables more objective evaluation of different ranking retrieval algorithms than a constant classification threshold. For example, if a result set has the maximum lift factor for $r = 5$ this implies that to achieve the highest precision classification only pictures with $r \leq 5$ have to be classified as True and all other pictures with $r > 5$ as False. In this study such discrimination typically included only the first 5–10 % of samples. This is in contrast to adaptation of classification for sensitivity where the classification threshold is arbitrarily set to a constant high value (90 % of the population in the experiment) to include as many true positive samples as possible. But this also results in lower precision because inevitably many false positives will also be included in the set. In reality using lift charts it is possible to choose what is more important – precision or sensitivity – and adapt the classification accordingly. The classifications were optimized for both goals with the lift charts segmented in 5 % intervals.

The order in which the returned documents were presented was disregarded because it is not possible to exactly establish the optimal order of documents, but only the partition of documents into the two classes. The aggregated experiment results are shown and elaborated in the next section.

3.3   Results and Discussion

The aggregated results of retrieval optimized for sensitivity and precision are presented in Table 1 and Table 2, respectively. Each table lists five retrieval performance measures (accuracy, precision, recall, fall-out and F-measure) for two lexical similarity algorithms which were used in 1-, 2- and 3-word queries. Additionally, results from the Table 1 and Table 2 are visualized in graphs in Figs. 5 and 6, respectively.

As can be seen in Table and in Fig. 5, with classifications optimized for sensitivity, multiword queries are better than single word queries in ranked retrieval of IAPS pictures: three word queries perform better than two-word queries, and they in turn are better than queries with only one keyword. However, the overall difference is small in some parameters.

The improvement between one and three keyword queries in accuracy is 17.22–21.11 % and even more in precision 24.23–26.19 % for both matching metrics. Compared to two-word search three keywords also show 8.56–12.22 % higher accuracy and 10.62–14.40 % precision. Likewise, two-word search performs 8.67–8.89 % better in accuracy and 11.79–13.62 % in precision than with one word. Recall is constantly high (92.48–95.75 %) since classification was optimized for sensitivity. Fall-out is the proportion of non-relevant documents that are retrieved out of all non-relevant documents and lower result is better. F-measure combines precision and recall as their weighed harmonic mean. Taken together these results could be interpreted that the retrieval is better in discrimination of non-relevant documents while a bit less capable in identification of relevant documents.

The data in Table 1 consistently shows a smaller difference in retrieval performance between individual matching algorithms than between single and multiword queries. This indicates that relatedness algorithms are less important than query size which strongly supports the need to implement query expansion and relevance feedback methods in affective multimedia databases search engines while the choice of lexical matching algorithm is less important.

Table 1. Aggregated results of classifications optimized for sensitivity.

Query	Matching	Accuracy	Precision	Recall	Fall-out	F-Measure
1-word	Exact	0.2911	0.1984	0.9556	0.0444	0.2328
	Approximate	0.2744	0.1912	0.9575	0.0425	0.2223
2-word	Exact	0.3778	0.3346	0.9330	0.0670	0.3536
	Approximate	0.3633	0.3091	0.9469	0.0531	0.3303
3-word	Exact	0.4633	0.4407	0.9248	0.0752	0.4515
	Approximate	0.4856	0.4531	0.9497	0.0503	0.4686

But if classification is optimized for precision, as presented in Table 2 and Fig. 6, approximate matching algorithms do not perform better than naïve exact matching. Incrementally adding more words to search query improves precision but reduces accuracy. Indeed, precision for 3-word queries in Table 2 was very high even reaching 100 % which should be viewed with caution as an artifact of the experimental dataset rather than a universal rule applying to all affective multimedia databases. These results could be explained with very low classification threshold in almost all instances of only 5 % (i.e. the most closely related 5 pictures to search query in $\dot{D}$). In such small sample both algorithms could perform well and accurately rank documents. Exact matching benefited from the choice of search keywords.

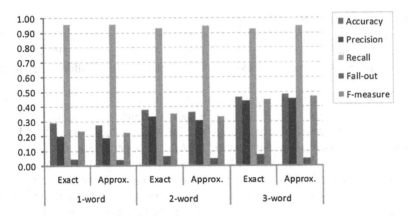

Fig. 5. Representation of performance measures in classifications optimized for sensitivity from Table 1.

Naïve string inclusion, as a representative of exact string matching algorithms, should represent the baseline algorithm that can be immediately used with all contemporary affective multimedia databases. As can be seen in Table 2 even this simple metrics can give sufficiently good results in some cases. Approximate matching algorithms may be regarded as improvements to affective multimedia databases which can be used only if the databases are upgraded with additional retrieval tools as was the case in this study.

Exact matching has high true positive rate, resulting in above average accuracy and recall, if search keywords are used for semantic annotation of pictures. This trend is particularly evident in precision optimization. The naïve algorithm requires from users to possess knowledge about the database keywords otherwise the retrieval performance will be modest. Since the experiment always used keywords present in randomly selected datasets $\mathbf{D}$ the exact matching could achieve relatively high quality of retrieval compared to approximate metrics.

In this experiment the Levenshtein algorithm represented the best retrieval method based purely on picture descriptions but only if retrieval is optimized for sensitivity. But if a small and accurate set of pictures needs to be retrieved from affective multimedia databases, and annotating vocabulary is known, exact matching and classification optimized for precision may the optimal solution.

In summary, the overall results imply that pictures in affective multimedia databases are insufficiently annotated yielding poor information retrieval performance scores. Accuracy, precision and recall are relatively low for all evaluated categories. It is evident that only one keyword per picture cannot convey higher meaning which results in higher false positive and false negative count. But, on the positive side, query expansion has been shown to always improve performance in precision, while accuracy will be better only for expanded queries in classification optimized for sensitivity. These results place constraints on limits of keyword retrieval in affective multimedia databases and can be used as a motivation in further research into more complex text-based retrieval methods.

Table 2. Aggregated results of classifications optimized for precision.

Query	Matching	Accuracy	Precision	Recall	Fall-out	F-Measure
1-word	Exact	0.79	0.7741	0.3349	0.6651	0.7668
	Approximate	0.8211	0.6185	0.2558	0.7442	0.6761
2-word	Exact	0.71	0.9292	0.1891	0.8109	0.8001
	Approximate	0.6944	0.8911	0.2117	0.7883	0.7749
3-word	Exact	0.6778	1	0.2528	0.7472	0.8071
	Approximate	0.6522	0.9899	0.1988	0.8012	0.7847

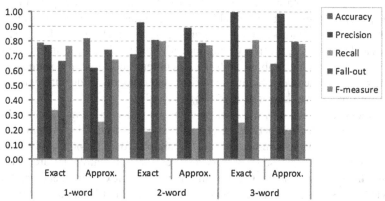

Fig. 6. Representation of performance measures in classifications optimized for precision from Table 2.

Since keywords are only lexically related semantically expressive picture descriptions can be implemented only by relying on more complex annotation models using semantic networks or ontologies. However, this would require significant modifications to existing affective multimedia databases by transforming their current keyword annotations to semantic concepts and adding references to higher knowledge structures. But as a consequence search queries could be expressed as ontological concepts and brought into connection with annotations using different semantic relatedness measures effectively enabling concept-based affective multimedia retrieval.

4   Related Work

Although keyword search is the most prevailing and explored method of multimedia retrieval little has so far been done to implement it in affective multimedia systems. Keyword search is the preferred type of retrieval in popular search engines (e.g. Google, Yahoo, Bing etc.) and internet picture repositories, such as Flickr, Google Picasa,

Photobucket, Shutterstock and many others, which are closely integrated with social networks and searched by hundreds of millions of users every day. Although almost all of these search engines support query-by-example, which is a content-based method with the goal of finding pictures visually similar to an example provided by the user, still the most dominant and accurate search method relies on metadata describing the picture content. A lot has been accomplished in text-based picture annotation and retrieval, regarding the development of various retrieval techniques, methods and algorithms, but research is still ongoing in many areas [28, 29].

The ontology-based multimedia retrieval, as a subtype of test-based search [30], has been intensively researched in recent years. Some of the most important multimedia ontologies specifically designed for annotation and retrieval of multimedia are TRECEVID [31], COMM [32] and LSCOM [33]. For a detailed overview of multimedia ontologies see [34]. Ontologies have been identified as one of the most promising improvements to metadata search [35]. In this paradigm metadata is mapped to concepts in an adequately expressive and formal knowledge ontological structure. Additionally, if possible, specific individuals of concepts are recognized and annotated. In the last step properties and relations between concepts and individuals of concepts in the knowledge base are used to infer facts about multimedia content and retrieve documents which match the posed search query. Semantic retrieval experiments using ontologies showed clear improvements over keyword-based approach [36].

Formal logical foundations with well-developed syntax and semantics, the ability to leverage knowledge models between high expressivity or formatting simplicity, standardized logical languages, numerous tools available for ontology engineering and knowledge retrieval are just some of important reasons why ontologies are suitable for modeling affective multimedia. On the other hand, complexity, efficiency and quality of knowledge bases are open research issues which have not been completely solved. General knowledge bases such as SUMO [37], ConceptNet [38] or DOLCE [39] are very good in describing a wide range of topics but because of their size and complexity require state-of-the-art computer tools. Furthermore, a single upper ontology cannot describe all aspects of the modeled world so additional smaller and specific ontologies are unusually needed to obtain a complete description of the domain. To a researcher working with affective multimedia these problems are an overhead if not solved by developers of affective multimedia databases. For practical purposes it should not be necessary, or in fact prudent, to demand from neuroscientists, psychologists or any other users to be acquainted with knowledge representation technologies. Furthermore, introduction of ontologies requires alterations to existing database models and creation of new formal annotations. The latter task seems particularly demanding and certainly demands expert input at least at the end of the ontology annotation process to manually approve each statement about stimuli semantics. If the process is not only centered on transferring informal semantics captured in existing keywords to formal ontologies but also aims to expand existing descriptions with additional statements about objects, events and other cognitive content in stimuli, then the creation of ontology-based affective multimedia databases is even more difficult and introduces more workload on the experts. In conclusion, although ontology-based multimedia knowledge models offer clear benefits, the existing free-text annotation schemes should be used to a maximum possible extent because any improvements can be immediately used to further affective multimedia databases.

Contemporary web search engines rely on integration of multimedia content and text information to achieve a more complete description of searched documents. This semantic search is focused on resolving the semantic gap problem (commonly referred to as "bridging the semantic gap" [40]) as in difference between low level descriptions of pictures based on picture elements and high level, i.e. semantically complete, understanding of scene with objects, events and their complex interactions.

5   Conclusion

The paper experimentally examined feasibility of using unsupervised keywords for image search in affective multimedia databases with the goals of providing recommendations for (1) development of multimedia retrieval tools, and (2) future research in improvement of databases' semantic models.

The experiment demonstrated the best performance for multi-word queries if classification was optimized for sensitivity. In fact, three-word queries achieved up to 26.19 % and 14.40 % better precision in ranked retrieval of IAPS pictures than one- and two-word queries, respectively. This would indicate that affective multimedia retrieval tools should primarily, and perhaps exclusively, use sensitivity optimized multi-word search.

Since affective multimedia databases predominantly use single word descriptions of documents the immediate search improvements can be accomplished by employing query expansion and relevance feedback methods, as well as with the selection of optimal lexical relatedness algorithms. Looking forward, keyword affective multimedia search has a significant potential if current semantic models are improved toward multi-word annotation with supervised vocabularies and employed with specialized intelligent retrieval tools.

Most inquiries to affective multimedia databases contain only a few terms requiring additional keywords to expand the original query. This process can be automatic or manual (as in the undertaken experiment). Semantically similar words, synonyms and discriminative words can be added to primary expert queries. Based on the reported experiment results, and taking into consideration the inherent difficulty in changing existing affective multimedia systems, it is necessary to use either query expansion or relevance feedback methods to enable multiple words searches. At the moment these methods are not adequately used with affective multimedia databases.

It is our recommendation that in the long-term all affective multimedia databases from the very beginning of their development should be integrated with sufficiently expressive knowledge bases for description of multimedia semantics to further improve the potential of their retrieval performance. Additionally, existing ontology resources useful for construction of annotation models and semantic data input should be extensively used in order to achieve better standardization among individual databases.

References

1. Horvat, M., Popović, S., Bogunović, N., Ćosić, K.: Tagging multimedia stimuli with ontologies. In: Proceedings of the 32nd International Convention MIPRO 2009: Computers in Technical Systems, pp. 203 – 208. Intelligent Systems, Opatija (2009)
2. Horvat, M., Popović, S., Ćosić, K.: Multimedia stimuli databases usage patterns: a survey report. In: Proceedings of the 36th International Convention MIPRO 2013: Computers in Technical Systems, pp. 993 – 997. Intelligent Systems, Opatija (2013)
3. Brave, S., Nass, C.: Emotion in human-computer interaction. In: The Human-Computer Interaction Handbook: Fundamentals, Evolving Technologies and Emerging Applications, pp. 81 – 96. CRC Press, Taylor & Francis, Florida (2002)
4. Coan, J.A., Allen, J.J.B.: The Handbook of Emotion Elicitation and Assessment. Oxford University Press Series in Affective Science. Oxford University Press, New York (2007)
5. Grandjean, D., Sander, D., Scherer, K.R.: Conscious emotional experience emerges as a function of multilevel, appraisal-driven response synchronization. Conscious. Cogn. **17**, 484–495 (2008)
6. Ćosić, K., Popović, S., Horvat, M., Kukolja, D., Dropuljić, B., Kovač, B., Jakovljević, M.: Computer-aided psychotherapy based on multimodal elicitation, estimation and regulation of emotion. Psychiatr. Danub. **25**, 340–346 (2013)
7. Lang, P.J., Bradley, M. M., Cuthbert, B.N.: International affective picture system (IAPS): Affective ratings of pictures and instruction manual. Technical report A – 8. University of Florida, Gainesville, FL (2008)
8. Lang, P. J., Bradley, M. M.: The International Affective Digitized Sounds (2nd Edition; IADS-2): affective ratings of sounds and instruction manual. Technical report B-3, University of Florida, Gainesville, FL (2007)
9. Gross, J.J., Levenson, R.W.: Emotion elicitation using films. Cogn. Emot. **9**(1), 87–108 (1995)
10. Villani, D., Riva, G.: Does interactive media enhance the management of stress? Suggestions from a controlled study. Cyberpsychology Behav. Soc. Netw. **15**(1), 24–30 (2012)
11. Marchewka, A., Żurawski, Ł., Jednoróg, K., Grabowska, A.: The nencki affective picture system (NAPS): introduction to a novel, standardized, wide-range, high-quality, realistic picture database. Behav. Res. Methods **46**(2), 596–610 (2014)
12. Bradley, M.M., Lang, P.J.: Measuring emotion: behavior, feeling and physiology. In: Lane, R., Nadel, L. (eds.) Cognitive Neuroscience of Emotion, pp. 242–276. Oxford University Press, New York (2000)
13. Dan-Glauser, E.S., Scherer, K.R.: The Geneva Affective PicturE Database (GAPED): A new 730 picture database focusing on valence and normative significance. Behav. Res. Methods **43**(2), 468–477 (2011)
14. Koelstra, S., Muhl, C., Soleymani, M., Lee, J.S., Yazdani, A., Ebrahimi, T., Pun, T., Nijholt, A., Patras, I.: DEAP: a database for emotion analysis; using physiological signals. IEEE Trans. Affect. Comput. **3**(1), 18–31 (2012)
15. Tottenham, N., Tanaka, J.W., Leon, A.C., McCarry, T., Nurse, M., Hare, T.A., Marcus, D.J., Westerlund, A., Casey, B.J., Nelson, C.: The NimStim set of facial expressions: judgments from untrained research participants. Psychiatry Res. **168**(3), 242–249 (2009)
16. The Paul Ekman Group, LLC. http://www.paulekman.com/product/pictures-of-facial-affect-pofa/
17. Bradley, M.M., Lang, P.J.: Affective norms for English words (ANEW): stimuli, instruction manual and affective ratings. Technical report C-1. Gainesville, FL. The Center for Research in Psychophysiology. University of Florida (1999)

18. Baccianella, S., Esuli, A., Sebastiani, F.: SentiWordNet 3.0: an enhanced lexical resource for sentiment analysis and opinion mining. In: Proceedings of LREC-10, 7th Conference on Language Resources and Evaluation, Valletta, MT, pp. 2200–2204 (2010)
19. Zeng, Z., Pantic, M., Roisman, G.I., Huang, T.S.: A survey of affect recognition methods: audio, visual, and spontaneous expressions. IEEE Trans. Pattern Anal. Mach. Intell. **31**(1), 39–58 (2009)
20. Gross, R.: Face databases. In: Li, S., Jain, A. (eds.) Handbook of Face Recognition. Springer-Verlag, Pitts-burgh (2005). The Robotics Inistitute, Carnegie Mellon University Forbes Avenue
21. Peter, C., Herbon, A.: Emotion representation and physiology assignments in digital systems. Interact. Comput. **18**(2), 139–170 (2006)
22. Russell, J.A.: A circumplex model of affect. J. Pers. Soc. Psychol. **39**, 1161–1178 (1980)
23. Mehrabian, A.: Pleasure-arousal-dominance: a general framework for describing and measuring individual differences in Temperament. Curr. Psychol. **14**(4), 261–292 (1996)
24. Bradley, M.M., Lang, P.J.: Measuring emotion: the self-assessment manikin and the semantic differential. J. Behav. Ther. Exp. Psychiatry **25**, 49–59 (1994)
25. Ekman, P.: Are there basic emotions? Psychol. Rev. **99**, 550–553 (1992)
26. Hliaoutakis, A., Varelas, G., Voutsakis, E., Petrakis, E.G., Milios, E.: Information retrieval by semantic similarity. Int. J. Semant. Web Inf. Syst. (IJSWIS) **2**(3), 55–73 (2006)
27. Navarro, G.: A guided tour to approximate string matching. ACM Comput. Surv. (CSUR) **33**(1), 31–88 (2001)
28. Hanbury, A.: A survey of methods for image annotation. J. Vis. Lang. Comput. **19**(5), 617–627 (2008)
29. Jacobs, P.S. (ed.): Text-Based Intelligent Systems: Current Research and Practice in Information Extraction and Retrieval. Psychology Press, New York (2014)
30. Hyvönen, E., Saarela, S., Styrman, A., Viljanen, K.: Ontology-based image retrieval. In: Proceedings of the XML Finland 2002 Conference, Helsinki, Finland, pp. 15–27 (2003)
31. Over, P., Awad, G.M., Fiscus, J., Antonishek, B., Michel, M., Smeaton, A.F., Kraaij, W., Quénot, G.: TRECVID 2010–An overview of the goals, tasks, data, evaluation mechanisms, and metrics (2011)
32. Franz, T., Troncy, R., Vacura, M.: The core ontology for multimedia. In: Multimedia Semantics: Metadata, Analysis and Interaction, pp. 145–161 (2011)
33. Naphade, M., Smith, J.R., Tesic, J., Chang, S.F., Hsu, W., Kennedy, L., Hauptmann, A., Curtis, J.: Large-scale concept ontology for multimedia. IEEE Multimedia **13**(3), 86–91 (2006)
34. Suárez-Figueroa, M.C., Atemezing, G.A., Corcho, O.: The landscape of multimedia ontologies in the last decade. Multimedia Tools Appl. **62**(2), 377–399 (2013)
35. Rahman, M.: Search engines going beyond keyword search: a survey. Int. J. Comput. Appl. **75**(17), 1–8 (2013)
36. Vallet, D., Fernández, M., Castells, P.: An ontology-based information retrieval model. In: Gómez-Pérez, A., Euzenat, J. (eds.) ESWC 2005. LNCS, vol. 3532, pp. 455–470. Springer, Heidelberg (2005)
37. Pease, A., Niles, I., Li, J.: The suggested upper merged ontology: a large ontology for the semantic web and its applications. In: Working Notes of the AAAI-2002 Workshop on Ontologies and the Semantic Web (2002)
38. Speer, R., Havasi, C.: ConceptNet 5: a large semantic network for relational knowledge. In: Kim, J., Gurevych, I. (eds.) The People's Web Meets NLP, pp. 161–176. Springer, Heidelberg (2013)

39. Gangemi, A., Guarino, N., Masolo, C., Oltramari, A., Schneider, L.: Sweetening ontologies with DOLCE. In: Gómez-Pérez, A., Benjamins, V. (eds.) EKAW 2002. LNCS (LNAI), vol. 2473, pp. 166–181. Springer, Heidelberg (2002)
40. Hare, J.S., Sinclair, P.A.S., Lewis, P.H., Martinez, K., Enser, P.G.B., Sandom, C.J.: Bridging the semantic gap in multimedia information retrieval: top-down and bottom-up approaches. In: Mastering the Gap: From Information Extraction to Semantic Representation/3rd European Semantic Web Conference, Budva (2006)

Data Driven Discovery of Attribute Dictionaries

Fei Chiang[1]([✉]), Periklis Andritsos[2], and Renée J. Miller[3]

[1] McMaster University, Hamilton, Canada
fchiang@mcmaster.ca
[2] University of Lausanne, Lausanne, Switzerland
Periklis.Andritsos@unil.ch
[3] University of Toronto, Toronto, Canada
miller@cs.toronto.edu

Abstract. Online product search engines such as Google and Yahoo shopping, rely on having extensive and complete product information to return accurate and timely search results. Given the expanding scope of products and updates to existing products, automated techniques are needed to ensure the underlying product dictionaries remain current and complete. Product search engines receive offers from merchants describing product specific attributes and characteristics. These offers normally contain structured attribute-value pairs, and unstructured (textual) descriptions describing product characteristics and features. For example, a laptop offer may contain attribute-value pairs such as "model-X42" and "RAM-8 GB", and a text description of the software, accessories, battery features, warranty, etc. Updating the product dictionaries using the textual descriptions is a more challenging task than using the attribute-value pairs since the relevant attribute values must first be extracted. This task becomes difficult since the text descriptions often do not follow a predefined format, and the data in the descriptions vary across different merchants and products. However, this information needs to be captured to ensure a comprehensive and complete product listing. In this paper, we present techniques that extract attribute values from textual product descriptions. We introduce an end-to-end framework that takes an input string record, and parses the tokens in a record to identify candidate attribute values. We then map these values to attributes. We take an information theoretic approach to identify groups of tokens that represent an attribute value. We demonstrate the accuracy and relevance of our approach using a variety of real data sets.

Keywords: Information extraction · Clustering · Dictionaries

1 Introduction

Product search engines provide a one-stop shop for users to compare and purchase products among the vast number of online merchants and the increasing number of products available. It is estimated that in 2013, approximately

R.J. Miller—Supported by NSERC BIN.

N.T. Nguyen et al. (Eds.): TCCI XXI, LNCS 9630, pp. 69–96, 2016.
DOI: 10.1007/978-3-662-49521-6_4

81 % of consumers performed online price comparison searches before making a purchase [2]. For these product search engines to be effective, comprehensive product listings are needed that accurately describe prices, attributes, and inventory. For example, a television can be described by a set of attributes such as manufacturer, model, size, and resolution. As manufacturers create new models, and enhance existing models with new features, this information is conveyed to product search engines, which must continually expand the breadth and depth of their product listings to keep pace with market dynamics. Given the expanding scope of products and updates to existing products, maintaining accurate product listings has become increasingly challenging.

Product information is normally sent via offers that consist of attribute-value pairs and textual descriptions. Attribute-value pairs describe product specific characteristics where the merchant has already associated a particular attribute to a particular value. For example, in the domain of television products, we may have attribute-value pairs such as "manufacturer-Samsung" and "model-LN32C450". As different merchants may use different attribute names and value formats to represent the same product characteristic (for example, an attribute called 'model' exists in the search engine product listing while the offer uses 'product-line'), the task of updating the product listings requires identifying the correct attribute correspondences between the product and the offers.

```
r₁: Samsung LN-52A650 52in 1080p 60Hz wide LCD screen
r₂: Mitsubishi WD65835 65" 1920x1080 Projection LCD
    TV wide screen
r₃: Sony XBR 1080p 32" LCD HDTV 120Hz
r₄: Samsung LN32C450 32" 720p 60Hz LCD HDTV
r₅: Sony BRAVIA KDL46HX850 120Hz 46-Inch 1080p 3D
    LED screen Internet TV
```

Fig. 1. Sample television data

Offers also contain unstructured text describing product characteristics that are not captured via the attribute-value pairs. (In fact, some offers may consist solely of a textual description.) Capturing the information in the text records is essential towards developing comprehensive and complete product listings. For example, consider the sample text records (extracted from offers) describing televisions, shown in Fig. 1. To update the product listings with this information, we need to detect parts of the text that represent properties such as manufacturer, model, size, refresh frequency, etc. Identifying which values correspond to which attribute is not necessarily clear, as we are not explicitly given the attribute-value mapping. Manually identifying the inherent structure and appropriate attribute values in such records is a laborious task requiring highly specific domain knowledge. Users would need to identify the desired attributes, and have knowledge of the syntax and semantics of these attributes. The specifications of a bicycle (e.g., frame, type) may be different from the specifications of a digital camera (e.g., memory, zoom).

Some offerings may include HTML tags and these can be used to extract attribute information. However, these tags may not always be correct, and different vendors may send inconsistent information causing inconsistencies among tags. Furthermore, not all data feeds have structured attribute-value pairs or tags available. Hence, extraction from textual product offers is needed. Current techniques rely on identifying basic attributes, such as manufacturer, and having a domain expert manually identify the remaining descriptive attributes (e.g., memory size). To extract values for attributes that follow a specific pattern (e.g., television refresh frequency) regular expressions can be used. However, for categorical attributes with no clear distinguishing formatting, such as TV manufacturer and model, reference tables, called *dictionaries*, are most useful [9,25]. Having an accurate and complete set of attribute dictionaries enables search applications to augment their product listings with a list of valid attribute values, and return more accurate query results. For example, a query for 'Sony' TVs with '1080 p' resolution can be answered quickly by referencing the manufacturer and resolution dictionaries for the matching values, and using indices to retrieve the matching records. Accurate dictionaries also provide a reference list of values for the attribute domain (e.g., all TV manufacturers). In some domains, standard tables such as states, and postal codes are readily available. Still, for domains such as televisions or children's clothing, finding a single authoritative, accurate, complete and correct source, listing all manufacturers, TV lines, or clothing lines is difficult.

Manual discovery and maintenance of dictionaries is laborious given the large number of possible attributes to describe a product and their corresponding (attribute) domain size. Previous work on dictionary discovery focused on expanding online product catalogs by using only the attribute-value pairs [19], and not the textual descriptions. In this paper, we take a data-driven approach for expanding and maintaining the product listings by focusing on dictionary discovery using the data *values*, and the relationships among these values. We do not assume that the schema information (of attributes) is known. We take as input a set of k attributes, and our goal is to produce a list of valid values for each of these k attributes.

We present techniques that generate and maintain a set of dictionaries, from a set of string records. Given a set of k attributes, our algorithms populate each of the k dictionaries. In order to discover which values belong to each of the dictionaries, we employ techniques that measure how much the individual values contribute to the information content of the original data. Each record is composed of a set of tokens or values (we use these terms interchangeably), as shown in Fig. 1. Tokens are substrings of a record. For simplicity in our examples, we assume that tokens are substrings that are separated by whitespace, but of course different tokenizers that account for punctuation, camel case or domain specific issues can also be used. We make the following contributions.

1. We present a data-driven approach to dictionary discovery that helps to expand and maintain online product listings. Our techniques group data values together such that each group represents a value in an attribute dictionary.

The grouping is based on an information theoretic approach that quantifies the information content among a group of values.

2. We describe a framework that captures the end to end dictionary discovery process. Given a set of string records, our techniques parse the records to identify candidate values. We then develop correspondences between a candidate value and an attribute dictionary. We realize that developing the correspondences can be highly domain specific, and we allow for domain user input (if available) to guide the correspondences.

3. We conduct a qualitative performance and comparative evaluation (against two existing systems) showing the effectiveness and efficiency of our techniques using a variety of real data sets.

Our paper is organized as follows. We discuss related work in Sect. 2. In Sect. 3, we give an overview of our dictionary discovery framework. Section 4 provides details of how attribute values are identified, and then refined using information-theoretic clustering techniques. Section 5 presents our algorithm for assigning candidate attribute values to attribute dictionaries. We present the evaluation of our techniques in Sect. 6, and conclude in Sect. 7.

2   Related Work

Previous work on segmentation has used machine learning techniques such as Hidden Markov Models (HMM) [6], Conditional Random Fields (CRF) [15,18, 28], and Bayesian networks [21]. Borkar et al. [6], present a nested HMM model that first learns the overall structure among elements in a record (e.g., state follows street in US addresses), and then learns the inner structure within each element. The algorithm assumes a given taxonomy of symbols to augment the learning process. Their primary focus is on discovering the inherent structure among the records and not on populating the attribute value lists.

In query segmentation [29] and keyword tagging of queries [25], research has focused on using generative models to maximize the probability that a candidate segmentation is the correct one. Most of these models either assume a fixed keyword order or assume a dictionary is given to facilitate the tagging process [10,17]. Sarkas et al. [25] investigate queries over structured Web tables. Given a keyword query q, their techniques annotate keywords by mapping each token to an attribute from a structured table. The annotation scheme assumes a knowledge base of structured tables is given and relies on a matching scheme to derive the annotations. In addition, they assume that the tokens are generated independently. Recent work by Lee et al., propose a probabilistic framework for attribute extraction given a large number of concepts, entities, isA relationships, a knowledge base, and various data sources [16]. Our work does not make independence assumptions, and we do not assume concepts are given nor that a full knowledge base exists.

A semi-automatic dictionary discovery tool is proposed by Godbole et al. [12] to build and refine dictionaries for text extraction. Their techniques assume the attributes, and initial words to start the search process are given. Words are grouped into the same dictionary based on a frequency, tf-idf (term frequency and inverse document frequency) based similarity model. Their results show low precision and recall numbers (in the range of 5–60 %). Our algorithms apply information theoretic techniques that capture both the frequency and information content between values, leading to higher qualitative results.

Nguyen et al. focused on expanding online product catalogs by using the attribute-value pairs and not the textual descriptions in online offers [19]. One aspect of the problem focused on deriving attribute correspondences between attributes extracted from the merchant offer and attributes from the catalog, since attributes describing the same feature may have different names. A second problem focused on merging attributes from multiple offers (that describe the same product) to create a new entry in the product catalog. Their solutions focus on using the structured attribute-value pairs, and not the unstructured text that we use. In addition, they assumed knowledge of key attributes (that uniquely identify a product), product attribute-value pairs, and the catalog schema.

Discovery of attributes in relational databases considers the problem of automatically clustering columns into attributes [31]. These techniques compute various statistical measures between all pairs of columns within the database, and determine positive and negative relationships between certain pairs of columns. Different measures of similarity (name, value or distribution similarity) are used to identify which columns are best to cluster to form an attribute value. In our work, we cluster data values (with no pre-defined associations to attributes) using information theoretic techniques to identify the most appropriate attribute. We intend to compare such statistical measures with our information theoretic measures.

Recent work in attribute extraction over Web sources mine for sentiment keywords in product reviews [3], and propose CRF based techniques using data sources such as Wikipedia [5], and top-k based websites (e.g., the top-k most influential people in a given year) [32]. Zhang et al. [32] propose an extraction algorithm using a part-of-speech (POS) tagger and a CRF training model. While their techniques leverage the inherent natural language that may exist in the descriptions of most top-k sites, the techniques are less adaptable to machine generated attribute descriptions where few POS tokens may exist.

Roy et al. study the dictionary refinement problem that aims to improve the precision and recall scores of existing dictionaries by minimizing the number of false positive entries and maximizing the number of true positive entries [23]. The problem is modelled as an optimization function to maximize the F-score, and a polynomial-time algorithm is given assuming a set of constraints to restrict the number of entries that can be removed. In our work, we work directly with textual descriptions to extract attribute lists. However, the proposed refinement techniques can be applied in a post-processing step once an initial set of dictionaries is generated.

Chaturvedi et al. propose a greedy based algorithm that identifies structurally and semantically similar patterns from text [7]. Their techniques assume that a domain expert or lexicon is available to manually convert semantically similar tokens to a standard representation. In our work, we do not assume a domain expert exists to identify related keywords, and we use information-theoretic clustering to identify related tokens in text.

3   Overview of Framework

Returning accurate product search results relies on having timely and accurate product information. Attribute dictionaries, which provide a reference listing of valid attribute values for a given domain, are a vital resource for efficient and effective product search. The dictionaries however, must be generated and maintained, as incoming product offers are received from merchants (Fig. 2). In this section, we describe our framework for automatically generating and maintaining a set of dictionaries from textual descriptions in online product offers. For a given offer, we assume that the text description is represented as a string record containing a set of tokens. For a set of products from the same category (e.g., televisions), we have a set of records R. Given a set of k attributes from the product listing, our goal is to populate the k attribute dictionaries using the tokens in R. We assume that the given k attributes (from the user) reflect the desired attribute lists. Our framework does not automatically generate finer granularity dictionaries from the given k attributes, as this involves assessing user intent, where learning techniques may be most relevant.

Figure 3 illustrates our dictionary discovery framework consisting of modules for parsing the string records, identifying correlated values that describe an attribute, and mapping candidate values to attributes. Each attribute-value mapping will augment the list of values for that attribute dictionary. We define a record $r \in R$ to be a single string record, $i.e.$, r is composed of a set of tokens t_i. Tokens are substrings of a record and the maximum value of i is the number

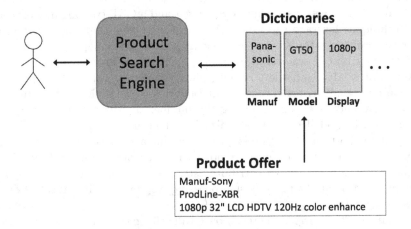

Fig. 2. Product search overview

of tokens in r. We assume the tokens t_i are substrings, separated by whitespace. We define a *segment* to be an ordered set of tokens t_i (we will relax this condition later).

Given a set of records R, describing products for a given domain (e.g., televisions), our first step is to identify segments that are potential attribute values, to be populated in the attribute dictionaries. For example, given a set of records (each describing a specific television), we may have a set of tokens 'wide' 'color' 'enhancement' occurring in this order. If these tokens co-occur frequently in this order, this provides evidence that these tokens are representative of an attribute value (in this case an image processing feature). We want to find all such segments as these will form the set of candidate attribute values. To do this, we construct a model containing all sufficiently frequent segments in R. By adding these segments to our model, we can think of removing them from the data R, and reducing the size of R. Our goal is to find segments that provide a minimal size model and a minimal data representation R. This notion can be quantified using the *Minimum Description Length (MDL) Principle* [11,22]. We apply the MDL Principle and use this intuition to find sufficiently frequent segments in the *segment identification* module. Note that we do not apply frequent item set counting techniques [4] since we are looking for sequences of tokens, and frequent itemset techniques do not consider order. Secondly, to limit the space of segments considered, we evaluate segments up to a fixed length. We choose the MDL principle for its flexibility as it allows us to succinctly model input token sequences across different domains by minimizing an appropriate objective function.

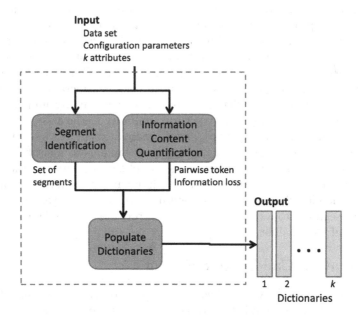

Fig. 3. Dictionary discovery framework

Our records are highly heterogeneous; with no predefined formatting nor specified order of attribute values, and not all the attributes may be described (i.e., missing values). MDL provides a concise and efficient representation of the data, allowing us to capture these subtle inconsistencies. For example, we may have one record containing tokens 'wide screen LED', another record containing 'LED wide screen', and another containing 'wide LED screen'. All these variations likely refer to the same attribute 'screen type', and identifying this co-occurrence relationship is difficult when the tokens do not consistently occur in the same sequential order. In our *information content quantification* module, we propose new techniques to compute the strength of co-occurrence between sets of tokens, regardless of their order of co-occurrence. We use an information theoretic measure, *information loss*, to assess how strongly two sets of tokens are related in the given data set. We use this information to modify the identified segments, by either removing tokens or adding new tokens. Previous techniques, such as Conditional Random Fields (CRF) have not considered this issue, and have relied on training data to provide the necessary format and structure of an attribute value [6,24]. To determine how a segment should be refined, we need to measure the relationship between two tokens. We use information content, as this measure considers the co-occurrence frequency between tokens, and is invariant to the order in which the tokens occur in the record. By doing this, we are able to compensate for the inherent heterogeneity, and identify segments consisting of tokens that appear in different orders in a record.

In the final module, *populate dictionaries*, with a set of refined segments in hand, we map each segment (from a record r) to one of the k attribute dictionaries. We do not assume that we know the mapping of values to attributes, only that k attributes exist. We propose a semi-supervised mapping algorithm, where we assume a user provides initial seed values to each dictionary, by mapping some of the refined segments to a dictionary. This provides us with a starting point from which to populate each of the k dictionaries, and enables us to produce relevant dictionaries according to a user's application needs. Our mapping algorithm first applies an elimination process, to remove from consideration, all segments in a record r that match a dictionary entry. The remaining unmatched segments in r are evaluated based on structural similarity to the existing dictionary entries. For example, 'win xp pro' and 'windows xp prof' are structurally similar. We use approximate string matching techniques to identify such pairings. We assume the attribute dictionaries model the regularity in the dataset. The amount of redundancy captured by the dictionaries is quantified again using the MDL Principle, which will help us determine whether it is worthwhile to add a particular segment to a dictionary. Next, we provide a more detailed description of each of the modules.

4   Generating Candidate Segments

Given a set of records R, where each record $r \in R$ describes a set of attributes for a product item, our goal is to extract the attribute values and map them to the corresponding attribute dictionary.

Example 1. Figure 1 shows five records, each describing a television and its attributes. Automatically extracting these product attributes is difficult since the records are heterogeneous and do not conform to a common structure. For example, the TV size does not always immediately follow the TV model (as in r_5) and values describing the television type do not always occur consecutively as in 'wide' and 'screen' which are consecutive in r_2 but not r_1. Given these records R, we would like to extract values for attributes describing a television, such as manufacturer, model, size, resolution, type, and refresh frequency.

We provide further details of our extraction process next.

4.1   Segment Identification

Recall that a segment is an ordered set of tokens from a record r. Given a set of records R, we want to extract segments that represent a product attribute. For example, in Fig. 1, we want to identify '1080p' and 'wide screen' as two different attribute values, for display resolution and type, respectively. We are interested in identifying a set of sequential tokens that may represent an attribute value, and we will consider mapping values to attribute dictionaries later (in Sect. 5). We identify segments by searching for sets of sequential tokens that co-occur frequently. Our intuition is that tokens that co-occur frequently provide evidence that they are related and represent a potential attribute value. We consider the set of records R as our data set. We preprocess the data to compute the frequency of candidate segments ranging in length from one token up to n tokens, for a given typically small n. That is, for each record $r \in R$, we sequentially scan the tokens and consider segments of increasing size. We store these candidate segments along with their frequency of occurrence in a *frequency table*.

Given the number of segments of varying sizes and frequencies, we need to determine which segments and their frequencies qualify as an attribute value (and consequently belong to a dictionary). Our approach involves translating the notion of frequently occurring segments into one of capturing the regularity in the data. We can associate a cost to represent R, say the number of bits needed to represent all the characters. If we are able to extract a frequently occurring segment s from R, we can reduce the representation cost of R by the cost to represent s, times the frequency of s. We can consider frequently occurring s as being *regular* in R. We build a model D that captures all such s, and models the regularity in R. To determine which segments s qualify to belong to D, we will use the MDL principle to guide us, a principled way to identify the model that best compresses the data. The MDL Principle states that the best model D is the one which has the minimal representation cost, and the minimal cost to represent the data given the model. For example, we can define a representation cost for the model D, $L(D)$, and a representation cost of the data given the model $L(R|D)$. We seek to add segment s_1 rather than segment s_2, if by adding s_1, the representation cost is less than if we added s_2, that is, $(L_{s_1}(D) + L_{s_1}(R|D)) < (L_{s_2}(D) + L_{s_2}(R|D))$. Our model D is a model of our attribute dictionaries. As attribute dictionaries are often constrained by performance requirements to provide fast search, our model D is necessarily constrained to contain meaningful segments s that offer the most bang for the buck. We apply a greedy approach, and consider segments in the order of decreasing size, to find non-spurious segments that reduce the description length cost the most.

The MDL principle states that the best model D for a data set is the one that leads to the best compression of the data. In other words, the model that is able to capture the most regularity in the data, is the one that leads to the best compression. We define the size (length) of a model D, $L(D)$ as $\sum_{t \in s} log_2(|t|)$, the sum of the log of the number of characters in each segment s, where t is a token in s, and $|t|$ is the number of characters in t. We define the initial size (length) of the data R given the model D, $L(R|D)$ as $log_2(|R|)$ where $|R|$ is the number of characters in all tokens in R. We can then compute a total *description length* $(DL) = L(D) + L(R|D)$, which is the total representation cost of the model, and the data given the model. Our goal is to identify s that can be added to D such that DL is minimized.

We begin with an empty D, and hence, $L(D) = 0$. When we consider adding a candidate segment s to D, we compute how the description length is affected. That is, we add the representation cost of s, $\sum_{t \in s} log_2(|t|)$, to $L(D)$, and reduce $L(R|D)$ by $log_2(f_s \cdot \sum_{t \in s} |t|)$, where f_s is the frequency of occurrence of s. We do this since the data that was in R is now captured in D. We only add s to D if it reduces DL.

Example 2. Consider Fig. 1. We start with an empty model D, and $L(D) = 0$. In this dataset, there are three occurrences of tokens of length 2 (two occurrences of TV and one of 3D), eight occurrences of tokens of length 3 and so on up to tokens of length 10. This gives a description length of about 98 assuming an empty model. We order our segments in decreasing size, and consider each segment in turn. Suppose we consider adding the segment 'LCD HDTV' to D, we get $L(D) = log_2 3 + log_2 4 \approx 3.58$, and reduce $L(R|D)$ by $log_2(2*(7)) \approx 3.81$. Since this segment reduces the current DL, it would be added to D. Alternatively, if we consider adding the segment 'Internet TV' to D, we get $L(D) = log_2 8 + log_2 2 = 4$, and reduce $L(R|D)$ by the same amount. Since there is no reduction in the DL value, we do not add this segment to D. We evaluate each segment in turn, and add only those segments to D which give a reduction in the DL value.

We have only considered segments consisting of sequential tokens thus far. We did this so as to not lose the order in which the tokens naturally occurred in a record. In addition, we wanted to select segments with tokens that occur frequently together indicating they are correlated. The MDL Principle allowed us to determine a sufficient frequency through the description length cost function. In the next section, we refine our segments to consider cases where attribute values may not involve sequential tokens. We apply information theoretic clustering techniques, which are invariant to the order of occurrence. We describe this process next.

4.2   Information Quantification

In this section, we describe our novel information theoretic approach that allows us to modify our segments to include attribute values involving tokens that may not occur sequentially in the record r. For example, a record may contain the tokens 'panasonic 3D viewer image panel neoplasma' and we have already identified the segment '3D viewer'. However, a more complete attribute value would include 'image' to form the segment '3D image viewer'. We apply information-theoretic clustering to understand the associations between sets of tokens that co-occur naturally in R, and to avoid adding spurious segments to our dictionaries. We check whether any of the newly created segments can be added to D to produce a more concise model.

We want to determine the strength of co-occurrence between two sets of tokens s and s'. For sets that share a high information content, we can add s' to s. For overlapping sets that do not share a high information content, we can remove s' from s. We evaluate tokens based on the amount of information content they share with other tokens [20]. Specifically, we use entropy, $H(V)$, which measures the uncertainty in the values V. The lower the entropy, the less uncertainty exists in the values. Conditional entropy, $H(V|R)$ is the uncertainty in the values given information on which record $r \in R$ the value belongs to. Then $H(V)$ and $H(V|R)$ are used to define mutual information, $I(V; R) = H(V) - H(V|R)$, which quantifies the amount of dependence between R and V. The relation $I(V; R)$ is symmetric, non-negative and equals zero if and only if R and V are independent.

To enumerate interesting combinations of tokens and evaluate their information content requires careful planning to avoid high computational complexity. The use of clustering techniques allows us to adopt a bag of words model where we can analyze token values, without worrying about the order in which these values occur in a record. We use mutual information to capture the association between sets of tokens that appear together (in any order) within a record. We employ a greedy hierarchical clustering approach, called *Agglomerative Information Bottleneck (AIB)*, [26], that starts from the most informative sets of tokens (high information content) and constructs clusters of naturally co-occurring tokens. Suppose we have two tokens u and v. If u and v co-occur exclusively together, then if we cluster u and v together, we do not lose any information since they are strictly dependent on each other. However, if u occurs with another token w (and v does not), then by clustering u and v, we incur some *information loss* since u also shares information with w. Now suppose c_x, c_y are two clusters of tokens. If we merge c_x and c_y, this indicates that there is a minimal information loss. In other words, the tokens are sufficiently correlated such that merging them incurs minimal information loss. This information loss is given by $\delta I(c_x; c_y) = I(C_l; R) - I(C_{l-1}; R)$, as the algorithm moves from a clustering of cardinality l to a clustering of cardinality $l - 1$. Intuitively, at each step, AIB merges two clusters that will incur the smallest δI.

Example 3. Figure 4 shows the dendrogram of successive merges, showing the initial values, intermediate clusters, and finally one large cluster, for the tokens

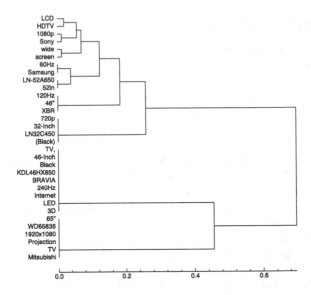

Fig. 4. Dendrogram of the data in Fig. 1

in Fig. 1. Examining the merges from left to right in Fig. 4, we see how information (x-axis) is lost as individual values are merged into sub-clusters. For example, values 'TV' and 'Mitsubishi' are merged with $\delta I = 0$ as they appear exclusively together. This is expected as they only appear once together in the original data set. If we consider frequently occurring tokens, e.g., 'wide' and 'screen', we see that the information loss during the merge is non-zero but still smaller than other values. Indeed, these two values appear together in the data records, except for record r_5, where the value 'screen' appears on its own. The latter is the record that introduces the corresponding loss of information.

We use the AIB algorithm and the dendrogram of sub-cluster merges to explore the amount of information that is contained in a data set. We use this to quantify the information content between sets of tokens. We consider modifying a segment s, such that the updated segment s' has minimal information loss and δI is within threshold bounds. We present our algorithms for merging and splitting segments next.

Refining the Segments. Given a set of segments and the dendrogram of tokens produced by the AIB algorithm, we want to determine if there are refinements to a segment s such that s contains a minimal amount of information loss. The segments produced thus far are based on having a sufficient frequency of occurrence without considering how much information is contained within the tokens. In this section, we describe how a segment s may be modified by removing tokens from s with less information content.

We introduce a vector representation of the dendrogram, and for the tokens in the original data set over the set of ordered merges $m_1, m_2, \ldots m_{n-1}$, where the order reflects the order in which the merge occurs as we scan the dendrogram from the tokens to the root. Keeping the sequence of merges from AIB, each token $v \in V$ is described by a binary $(n-1)$-vector M_v, where $M_v(i) = 1$ if and only if v participates in merge i and 0 otherwise.

Algorithm 1. FIND_MERGE(s)

Input: String of tokens s; Vectors $M_v, \forall v \in \mathbf{V}$; Vector I;
Output: First merge m of tokens in s in which all tokens of s participate; Information Loss
     value of merge m;
 1: Initialize vector M_{new} with $|M_v|$ zeros;
 2: Split s in n_s tokens $s^1, s^2, \ldots, s^{n_s}$; { s contains > 1 tokens }
 3: **for** $i = 1$ to n_s **do**
 4:     $M_{new} = M_{new} + M_{s^i}$;
 5: **end for**
 6: **for** $j = 1$ to $|M_{new}|$ **do**
 7:     **if** $M_{new}(j) = n_s$ **then**
 8:        **return** $\left(j, I(j) \right)$;
 9:     **end if**
10: **end for**

This representation is used in Algorithm 1, FIND_MERGE, to find the merge point where all the tokens of a segment are first merged. The algorithm uses vectors M_v, as well as a vector I that stores the information loss of each merge m_i, $1 \le i \le (n-1)$. A new vector M_{new} accumulates the vectors of the tokens in s (lines $3--5$). Since all M_v vectors are binary, it finds the first index of M_{new} whose value is equal to the number of tokens in s (lines 6–9). This merge together with its loss of information are returned (line 8).

If a segment contains a subset of tokens with higher information content, i.e., less information loss than the entire segment, we would like to extract this subset, and use it as the updated segment. To do this, we introduce Algorithm 2, called SPLIT. Given a threshold, τ_{split}, SPLIT splits a segment into subsets of tokens whose merge is the maximum information loss below this threshold. It uses FIND_MERGE to assess the content of all subsets and inserts them into a maximum heap, in decreasing order of their information loss (lines 1–6). While the heap is not empty, all subsets are removed. If their information loss is less than the threshold, they are included in the resulting sub-segments (lines 7–18). This process allows us to remove tokens that may not be relevant to the attribute domain at hand.

Example 4. Continuing our running example, there are $n = 32$ tokens and, therefore, $n - 1 = 31$ total merges as we scan the dendrogram from the leaf level towards the root. Considering $\tau_{split} = 0.04$ (indicated on the x-axis of the dendrogram), and the segment 'LCD TV wide screen' identified by MDL, we see that it should be split into sub-segments 'LCD', 'wide screen' and 'TV' as these are the only segments whose tokens participate in merges below τ_{split}.

Algorithm 2. SPLIT(*seg*)

Input: Segment *seg*; split threshold τ_{split};

Output: $\mathbf{G} = \Big\{ \{seg_1\}, \{seg_2\}, \dots, \{seg_g\} \Big\}$

    s.t. $seg_i, 1 \le i \le g$ are non overlapping subsets of G;

1: $S \leftarrow$ set of tokens in *seg*;

2: $Q \leftarrow$ max-heap storing segments according to the information loss of their token merges;

3: **for all** $\{s\} \in 2^S$ **do**

4:     [IL(s), MERGE(s)] $\leftarrow$ FIND_MERGE(s);

5:     Insert pair [IL(s), MERGE(s)] in Q;

6: **end for**

7: $i \leftarrow 1$;

8: **while** ($size(Q) > 0$) **and** ($S \ne \emptyset$) **do**

9:     $[il, m] \leftarrow$ pop top element from Q;

10:    **if** $il \le \tau_{split}$ **then**

11:      *candidate* $\leftarrow$ segment of top element stored in Q;

12:      **if** *candidate* $\subseteq S$ **then**

13:        $seq_i \leftarrow candidate$;

14:        $S = S - \{\text{tokens of } seq_i\}$

15:        $i \leftarrow i + 1$;

16:      **end if**

17:    **end if**

18: **end while**

---

5   Populating the Dictionaries

In this section, we describe our solution to populate each of the k dictionaries. Our final task is to populate the dictionaries $d_i, i = 1 \dots k$ with segments $s \in D$. Consider the set of records R, and for an $r \in R$ containing tokens that may match segments in D, we want to define mappings between s and d_i. Our problem is to map segments that describe the same product attribute all to the same dictionary. We approach this problem using the idea that the k dictionaries will collectively model our data each one of them is able to model the regularity in the data. That is, each attribute dictionary will contribute towards reducing the representation cost of the data if it contains segments describing the same attribute. If a segment s is mapped to the wrong dictionary, then (assuming that a record can only be mapped once to a specific d_i) this causes another segment also to be misplaced. For example, if the segment '60 Hz' was mapped (incorrectly) to the manufacturer dictionary, then for the record 'Sony Bravia 60 Hz ...', the token 'Sony' would also be misplaced (i.e., mapped to a dictionary that does not represent manufacturer), or not mapped at all to any dictionary.

Definition 1. *We define an **s-i mapping, map(s,i)** between a segment s and the dictionary d_i, $i = \{1, \dots, k\}$ to indicate that s belongs in dictionary d_i.*

As a starting point to populate the k dictionaries, and to allow users to customize these dictionaries to their application requirements, we allow users to define s–i mappings for the segments in D. This creates an initial set of entries for each of the attribute dictionaries, and allows us to produce more relevant dictionaries based on the user's starting preferences. We do not assume that all the segments in D have participated in an s-i mapping, hence, some dictionaries may have more initial entries than other dictionaries or no initial entries at all.

Definition 2. *Given a record r from R, we say that r **matches** a segment $s_i = \{t_1, \ldots, t_n\}$, if r contains distinct tokens $\{t'_1, \ldots, t'_n\}$, occurring in the same order as s_i, and $t_j = t'_j$, for each $j \in \{1, \ldots, n\}$.*

For every set of tokens in r that matches a segment in d_i, we replace these tokens with the dictionary identifer d_i. We do this to exclude these tokens from the mapping process since they have already been mapped by the user.

Definition 3. *If a record r contains l dictionary identifiers, these identifiers partition the set of tokens in r into $l + 1$ **sub-lists** of tokens.*   •

For example, $r_1 = \{t'_1, t'_2, d_4, t'_5, \ldots, t'_{n-1}, t'_n\}$ contains d_4, which partitions the tokens into two sub-lists $l_1 = \{t'_1, t'_2\}$ and $l_2 = \{t'_5, \ldots, t'_{n-1}, t'_n\}$. Our goal is to map these tokens in r_1 to a dictionary. We partition the set of tokens into sub-lists, and work with the tokens in each sub-list to find the most suitable dictionary. We do this since we assume that the identified segments act as delimiters for attribute values.

Definition 4. *An **assignment** $c_r = \{map(s_1, 1), map(s_2, 2), \ldots, map(s_v, k)\}$ for a record r is a set of s-i mappings that specify, for each segment $s \in D$ that matches in r, which dictionary (d_i) to which s will be added.*

There are a large number of possible ways to map each segment to one of the k dictionaries. To reduce this space of possibilities, we define the *validity* property that each assignment must satisfy.

Definition 5. *An assignment c_r is considered **valid** if there is at most one s-i mapping containing dictionary d_i.*

In other words, a record r can only contain at most one value from a dictionary d_i. For example, a record describing digital cameras cannot contain both Nikon and Canon as manufacturers. We will consider only valid assignments.

We develop a greedy mapping algorithm that sorts the records in increasing order of unmapped segments[1] and selects records with the fewest unmapped segments to map first. We use a list L of queues $L = (q_0, q_1, q_2 \ldots q_k)$ that contains records with unmapped but matched segments. The queue q_0 contains records with zero segments to map, q_1 contains records with one available segment to map, and so on. In addition, there is an incomplete queue, q_{incomp}, that contains records that continue to have available segments, after a segment fails to satisfy the membership criteria for a dictionary d_i. For these incomplete records, we could not find a suitable dictionary for the segments. Next, we describe the selection criteria we use to find a winning assignment.

---

[1] Since a record contains tokens that may match a segment, we can think of the records containing segments.

Example 5. Suppose D contains the list of segments {1080 p, LCD HDTV, 32",
Sony, 60 Hz, 720p}, and a user has defined an s-i mapping to associate segment
Sony to the manufacturer dictionary, and 32" to the size dictionary. Given the
records R in Fig. 1, record r_1 contains zero mapped segments (i.e., 'Sony and
32' do not appear in r_1) and two matched segments (i.e., '1080 p', '60 Hz'),
r_3 contains two mapped segments for the manufacturer and the television size
and two matched segments. Given D, we get the following queue list L:

q_1: r_2, r_5
q_2: r_1, r_3
q_3: r_4

5.1. Selecting a Winning Assignment

An assignment c_r for a record r specifies the segments that will populate the k
dictionaries. To find a winning assignment, we consider three selection criteria.

1. Each assignment c_r must be valid.
2. The assignment c_r must maximally reduce the total description length cost.
   We want to find segments to add to the dictionaries that capture the regularity
   in the data, and we model this via the description length.
3. If there is more than one assignment that satisfies the above two criteria, we
   select the assignment that maximizes the similarity between s and the entries
   in the corresponding dictionary d_i.

The first condition is to ensure that an assignment does not cause a record
to have more than one value from a dictionary. This condition must hold for
all records in R. Hence, when evaluating an assignment, we need to verify that
the segment to be added to d_i does not co-occur (in r) with any of the values
already in d_i. The second condition is to find segments that occur regularly in
the data. We use the MDL Principle, which allows us to exploit the regularity
in the data to compress the data. We use Ω to model the set of all dictionaries
$d_i, i = \{1, \ldots, k\}$. We seek the smallest description length $DL = L(\Omega) + L(R|\Omega)$.
If the initial model Ω is empty, then $L(\Omega) = 0$. Correspondingly, $L(R|\Omega)$ is equal
to the sum of the frequencies for each token in R. We initialize Ω to include the
initial entries in d_i, for $i = \{1, \ldots, k\}$. For each entry s in d_i, we add to $L(\Omega)$
the number of occurrences of s in R. Correspondingly, we reduce $L(R|\Omega)$ by the
frequency in which each token t (belonging to s) occurs in R in the same order
as it appears in s.

Example 6. Continuing our example, consider record r_4 with one mapped segment (television size), and three matched but unmapped segments. For the resolution and refresh frequency dictionaries, we add a frequency of 1 (segment
720 p occurs once in R), and a frequency of 2 (segment 60 Hz occurs twice in
R) to $L(\Omega)$. Consequently, we reduce $L(R|\Omega)$ by the same amount (since these
segments are of length one). For the segment $s =$ LCD HDTV, the tokens LCD
and HDTV each occur in R with frequency 4 and 2, respectively. In this case, we

increase $L(\Omega)$ by 2 (representing the frequency of occurrence of s), and reduce $L(R|\Omega)$ by $2 + 2 = 4$ for each occurrence of the (single) token in R.

If there is more than one assignment c_r that maximally reduces DL, then we consider the similarity between the segment s (from c_r) and the entries in the candidate dictionary d_i. We compare s and each entry in the candidate dictionary d_i, and for each pairing, we compute a similarity score (we currently use the Jaro similarity measure, but our techniques are agnostic to the similarity measure used). If this score is above a given threshold, we select c_r as the winning assignment, and add s to d_i. We use similarity as a tie-breaking mechanism to find segments that are most structurally consistent with the existing dictionary entries. For example, we may have a segment 4 GHz that is compared against entries from a computer processor dictionary that contain other processor speeds. If the minimum similarity threshold is not satisfied for all segments in c_r, we disregard this assignment, and consider a new candidate assignment. Once we find a winning assignment, we match the remaining records $r \in R$ with the latest entries in d_i, and update the queues in L. Further details of the mapping algorithm are given in Algorithm 3.

6   Evaluation

We conducted a qualitative, performance, and comparative evaluation of our techniques using real data sets. We wanted to investigate how our algorithms performed and scaled as we varied various parameters. In addition, we were interested in validating whether our algorithms produced accurate and complete dictionaries for the given domain, particularly against existing techniques. Our experiments were run using a Dell PowerEdge R310 with 64 GB memory, quad core Xeon X3470, 2.93 GHz, running Ubuntu 10.04.

Real Data Sets. We use seven real data sets. The first two data sets, namely *bikes* and *laptops* were gathered from the online merchant pricegrabber.com. The bikes data set is a collection of 3.2 k records, with six attributes, describing properties such as manufacturer, model, size, gender, type of bike, and accessories. These record values do not appear consistently in the record. For example, some of these attribute values may be missing, appear in a different order, and have inconsistent representations. The laptops data set consists of 13 k records, and eight attributes, describing properties such as manufacturer, product line, model, memory, and hard disk.

The next two data sets, *address* and *yellowpages*, contain, respectively, 120 and 275 k address strings scraped from online sources and yellowpages directory for small businesses. The *address* data set contains three attributes (street number, street name, street type) describing North American addresses. In the yellowpages data, each line contains an address string with five attributes describing the street number, street, city, state, and country. The final three data sets include *insurance, movies* and *Yelp* data. The insurance data set consists of 20 k records with 12 attributes describing insurance claims such as customer, type of claim, cause, and damages. The movies data set contains 8 k

---

Algorithm 3. MAP_SEGMENTS()

Input: Initialized d_i; data R; set of k attributes A, similarity threshold θ;
Output: Populated dictionaries $d_i, i = \{1 \ldots k\}$;

1: bestDL $\leftarrow \infty$; $best_{c_r} \leftarrow \emptyset$;
2: Replace segments in R, that belong to d_i, with d_i
3: Load R into $L \leftarrow (q_0, q_1, \ldots, q_k)$ based on the number of unmapped segments in the record
4: **while** (L non-empty) **do**
5:     $r \leftarrow$ pop(L); % the record with the fewest number of unmapped segments
6:     $B_r \leftarrow \{A\backslash \{\text{attributes marked in } r\}\}$;
7:     **if** $(B_r = \emptyset)$ **then**
8:         next; % get a new r
9:     **end if**
10:    Generate a candidate assignment, c_r, between B_r and segments in r
11:    **for** (each valid(c_r)) **do**
12:        $DL_{c_r} \leftarrow$ computeDL(c_r);
13:        **if** $(DL_{c_r} < \text{bestDL})$ **then**
14:            $best_{c_r} \leftarrow c_r$;
15:            bestDL $\leftarrow DL_{c_r}$
16:        **else if** $(DL_{c_r} = \text{bestDL})$ **then**
17:            **if** (maxSimilarity(c_r, d_i) $\geq \theta$) **then**
18:                $best_{c_r} \leftarrow c_r$;
19:            **end if**
20:        **end if**
21:    **end for**
22:    add segments in $best_{c_r}$ to corresponding d_i
23:    Update (mark segments in) R. Update L.
24: **end while**

---

records and 10 attributes describing films from Rotten Tomatoes. These film attributes include movie title, release day, year, season, studio, genre, and box office earnings. Finally, the Yelp data is part of the Yelp dataset challenge with string records for approximately 12 k businesses over 15 attributes, with tokens describing the business name, location, operating hours, category, type, rating, and price range [1].

Parameters. We use the Jaro-Winkler similarity measure [30] in the mapping algorithm for its ability to consider string length, token swaps, and the number of matching characters in the similarity score computation. Unless otherwise noted, for all experiments, we set the information loss value to 0.25, and the similarity value to 0.7. We evaluate the sensitivity of the information loss and similarity parameters by varying their values from 1 % to 100 % to determine the performance and qualitative effects. We found that adjusting these parameter values had little impact on the algorithmic running time, the performance remained steady across all tested values. We report the parameter sensitivity for the qualitative results in Sect. 6.2.

6.1   Quality Evaluation

We evaluate the quality of the generated dictionaries by computing the precision and recall of each attribute dictionary, across four data sets. For each dictionary, we compute precision and recall as follows:

$$precision = \frac{\#correctEntries}{\#returnedEntries}$$

$$recall = \frac{\#correctEntries}{\#entriesAttributeDomain}$$

To calculate the number of correct entries, we compare the values in each dictionary against the gold standard. The gold standard is based on the actual (true) attribute mapping of each value. That is, we manually review the returned entries, and validate using a set of online product websites whether the product attributes are categorized to the correct dictionary. If so, then they are marked as correct. For example, segment 'mountain' is categorized as a 'bikeType', and 'Huffy' is categorized in the 'manufacturer' dictionary. Similarly, we determine the total number of entries for an attribute domain by reviewing all values in the data set, and identifying (again, against a set of online bike and laptop sites) those values that represent one of our k attributes. The precision and recall results (averaged across all attribute dictionaries for each data set) are shown in Figs. 5 and 6, respectively. The precision results show that the more structured data sets, address and yellowpages, return the highest precision of 90 % or higher. The consistent formatting of attribute values helps the mapping algorithm to identify the correct dictionary. The less structured laptops data set showed an average precision of approximately 74 %. While the manufacturer, model, and product line dictionaries demonstrated fairly good precision of approximately 85 %, the dictionary for the memory attribute contained values from the processor speed attribute (and vice-versa), causing the precision for both dictionaries to be lower. However, correct values were identified for values such as 'core duo', and 'centrino pm' even though these values did not necessarily appear consecutively in the records. The ability to discover these non-sequential values accounted for an average of 21 % of the correct values across all the laptop dictionaries. Attribute dictionaries, with values that share a similar structure, such as laptop models (a60 vs. x60), processor types and speeds, and display types and sizes, benefitted from the similarity matching function in the mapping algorithm. Finally, the heterogeneous bikes data set showed an average precision of 60 % across all dictionaries, where the manufacturer, model, type, and size dictionaries showed the best result.

The recall results showed the same trend as the precision results, with the more structured data sets performing better, as expected. For the laptops data set, the dictionaries for model, product line, and processor type showed the best recall with results of 80 % or higher, showing that we are able to capture most of these technical specifications. Finally, the average recall across all the bikes dictionaries is 70 %, with the bike type and size showing the best results at 86 % and 100 % recall, respectively. As the regularity in the data improves, so do the precision and recall results. We note that these quality results are based on approximately 10 % of the input data serving as seeds in the dictionaries. That is, each of the dictionaries contained initial entries that helped to guide the mapping process. As part of our future work, we plan to quantify the relationship between the amount of input data and the quality of the dictionaries.

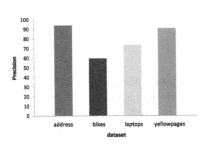

Fig. 5. Computed precision.   Fig. 6. Computed recall.

6.2 Parameter Sensitivity

Our algorithm contains a set of parameters that customize the dictionaries according to the user's application requirements. Specifically, we used the laptops data set, and evaluated the qualitative and performance impact as the values for the information loss, and similarity changed. We varied these parameter values from 1 % to 100 %. Our performance test showed that there was minimal impact on the running time, as the time (across both parameters) remained steady at about 6.5 min. We did observe a qualitative difference as shown in Fig. 7. As the information loss values increased, the precision decreased while the recall increased. If a user is willing to tolerate an increasing amount of information loss (that is, less information content between two values), then a given dictionary will contain lower information content among its values. This will cause more extraneous values to appear in the dictionary (lower precision), but an increased likelihood to capture all the domain values for that dictionary (higher recall). Figure 7 shows that the best information loss value is below 0.5 for users that are interested in achieving a balanced precision and recall rather than sacrificing one for the other. In future work, we intend to investigate how active learning techniques can help to minimize user effort, and to solicit user input only for the most necessary cases.

As we varied the similarity values, we observed a strong effect on dictionary quality. For increasing similarity values, the precision increases as more syntactically similar values are found. However, this decreases the scope of values that are returned in the dictionary (only the most similar values are found), causing the recall to decrease. Figure 7 shows that a similarity value between [0.7, 0.8] gives the best trade-off between precision and recall. At similarity value equal to 0.9, there was a large reduction in the number of returned dictionary entries, causing the recall to drop sharply.

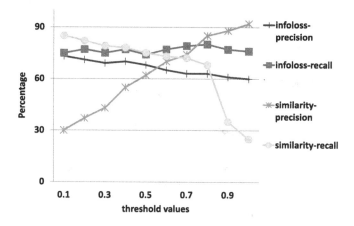

Fig. 7. Parameter sensitivity.

6.3   Performance Evaluation

We wanted to investigate the running time and scalability of our techniques. In particular, we measured the running time against the number of tuples, the number of dictionaries (attributes), and varying the number of domain values in the data set. We use the laptops data set, and duplicated the tuples to achieve a larger data set for testing.

Number of Records. In Fig. 8, we vary the number of records in the data from 50 k to 300 k, and measure the running time. The running time shows a linear progression. About 59 % of the running time is spent in the information quantification phase to compute the information loss among each pair of values. We are working towards improving the algorithm in this module so that only a subset of values need to be compared (instead of n^2 values), to compute the information loss. We anticipate that after this improvement, the total discovery time will be significantly reduced.

Number of Dictionaries. We vary the number of generated dictionaries from 2 up to 12. The total time remains under (a reasonable) 10 min even for 12 dictionaries. When a value is examined, and satisfies the criteria for a dictionary membership, then the value is mapped to that dictionary. Otherwise, the value is mapped to an incomplete list. For each dictionary, there is a fixed comparison time to perform the membership check. Hence, as the number of dictionaries increases, the number of membership checks increases, and we expect an increase in the total running time. We observe this gradual increase in Fig. 9 starting with 4 dictionaries. We believe the slight dip in running time between 2 and 4 dictionaries is due to a decreased number of values being mapped to the incomplete list (and consequently, an increased number of values mapped to the dictionaries), requiring less overhead to maintain the incomplete list.

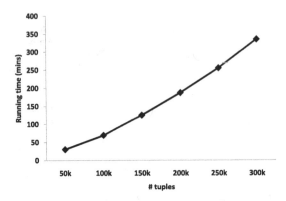

Fig. 8. Time vs. no. tuples

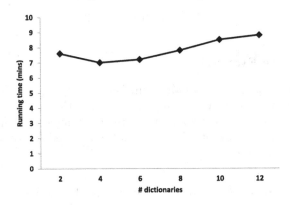

Fig. 9. Time vs. no. dictionaries

Number of Values. We studied the impact of increasing the number of domain values in the data set from about 100 to 550 values. Figure 10 shows that the discovery time scales with a steep linear curve. This is again primarily due to the computation of information loss in the information content quantification module. Since this module compares every pair of values and computes the information loss between these two values, an increase in the number of values, will cause an increase in the running time. However, the running times still remain under 7 min for up to 550 distinct values. We believe that these times can be further improved. We are exploring ways to do the value comparison, including using a threshold to identify the values to be compared, and using more efficient data structures rather than the vectors currently used.

6.4   Comparative Evaluation

We perform a comparative study to evaluate our algorithms against two existing techniques, Conditional Random Fields (CRF), and the Stanford Natural

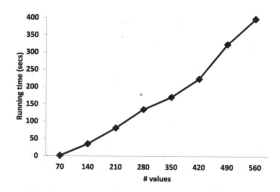

Fig. 10. Time vs. no. values

Language Parser (NLP) [14, 27]. For four data sets, we used the CRF implementation developed by Sunita Sarawagi [13], and trained the models using a subset of the data (approximately 10 %). Figure 11 and Fig. 12 shows the comparative precision and recall values. Our dictionary discovery algorithm achieved higher precision than CRF across the four data sets. Our techniques performed well with the address and yellowpages data, particularly in the yellowpages data set where we achieved 91 % precision versus 78 % in CRF. The CRF model could not distinguish between the street number and street names containing numeric values, such as '5th Ave'. For the bicycles data set, both techniques achieved similar precision of around 60 %. This score was caused by a lower precision value for the bike models dictionary. In the laptops data set, we achieved a significantly higher overall precision of 73 % over CRF's 57 %, due to our use of similarity measures that identify relevant groupings of values into segments. For example, our use of a similarity measure helps to associate segments such as 'win xp pro' and 'windows xp pro' into the same dictionary, whereas CRF does not consider approximate matching. In addition, by considering segments of non-sequential tokens, we can identify segments such as 'intel pentium iii' automatically (even though these tokens may not always occur consecutively in a record).

For recall, our algorithm had a few challenges in the address and yellowpages data sets. First, for street names that consisted of several tokens, the last token was identified as the street type. Similarly, in the yellowpages data set, the state was often miscategorized as the street type. These are both due to longer than expected street names. However, our techniques were able to achieve recall results that were 29 % and 23 % improvement over CRF, for the bikes and laptops data sets, respectively. The bikes and laptops data sets contain less structured records, with an increased amount of heterogeneity when compared to the address data records.

We evaluated our techniques against the Stanford NLP Parser (v3.2) using the insurance, movies, and Yelp data sets. This is a probabilistic parser that identifies grammatical structure in sentences and groups tokens according to parts-of-speech tagging, such as nouns, subjects and noun modifiers. We used these

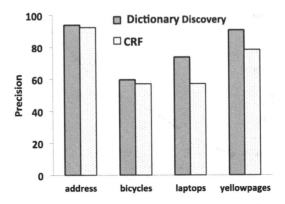

Fig. 11. Precision vs. CRF

tags to identify attribute values. Figures 13 and 14 show the precision and recall results, respectively. The precision and recall results show that our techniques consistently outperform the NLP parser by at least 20 %. The NLP parser segments the tokens according to sentence structure, often grouping adjectives with nouns. For example, in the movies data set *'Horror'* (film genre) is grouped with *'Monster'* (title). Similarly, in the insurance data set, the claim type (*'Auto'*) is grouped with damages (*'personal injury'*), leading to lower precision and recall scores. In the Yelp data set, the correct nouns returned by the NLP parser involved location information. The NLP parser incorrectly grouped data values from the business name (name attribute) with values from the remaining attributes, resulting in lower scores. In our techniques, we allow the user to provide initial training data (seeds) to guide the dictionary extraction and mapping process to achieve improved precision and recall. For these experiments, we use 10 % of the input data as seeds. Given the heterogeneity of these data sets, these results demonstrate the value that automated dictionary discovery provides to identify more complete lists. This is important for online vendors who aim to return as comprehensive results as possible to potential users.

6.5   Web Based Tool

We have implemented our algorithms into a dictionary discovery tool that offers an interactive web interface implemented using the jQuery UI Javascript library running on an Apache server [8]. Our tool allows a user to upload their own data, and control the thresholds of the qualitative measures. For example, if a user would like the data values to be matched based on distinctive data patterns, then a high similarity threshold should be used. Users define their schema by specifying a list of attributes, where a dictionary will be created and populated for each attribute. A priority ordering over the attributes can be specified by dragging and dropping the attribute to the desired ordered position, as shown in Fig. 15. Attributes at the top have highest priority, indicating that the algorithms

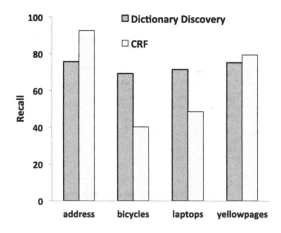

Fig. 12. Recall vs. CRF

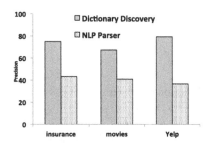

Fig. 13. Precision vs. NLP

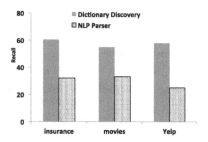

Fig. 14. Recall vs. NLP

will try to maximize the quality of these attribute dictionaries over attributes lower in the ordered list, by comparing the similarity values of these attributes first against the similarity threshold. In Fig. 15, we use a laptops dataset focusing on IBM Thinkpads and define eight attributes, with the Models attribute having the highest priority. If users would like to customize the dictionaries towards their application requirements, they can drag and drop the identified tokens (top right panel of Fig. 15) into the appropriate attribute boxes. Based on these inputs, we return a set of populated dictionaries. Our tool is available at http://dblab.cs. toronto.edu/dict.

7   Concluding Remarks

In this paper, we presented new algorithms for generating attribute dictionaries. Our techniques focused on using information theoretic techniques to identify values in the data with high information content, regardless of their order of occurrence. Our evaluation showed that we are able to discover dictionaries containing relevant values for the given attribute domain.

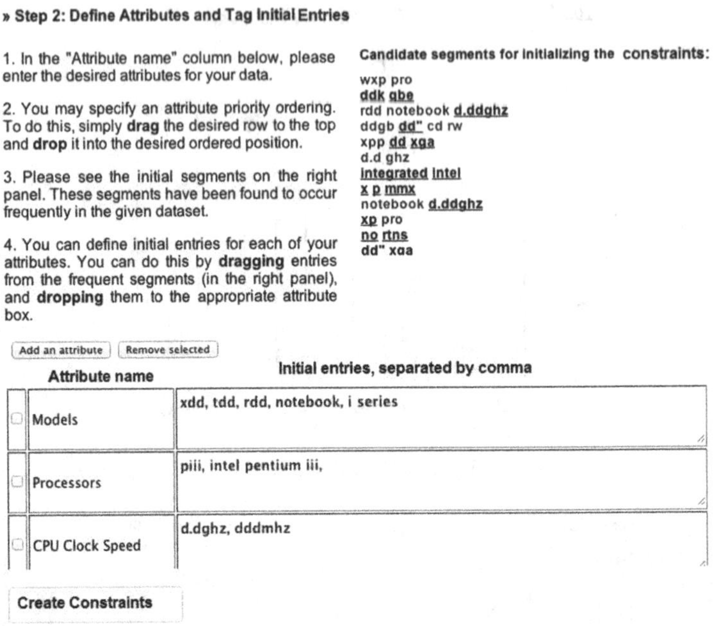

» **Step 2: Define Attributes and Tag Initial Entries**

1. In the "Attribute name" column below, please enter the desired attributes for your data.

2. You may specify an attribute priority ordering. To do this, simply **drag** the desired row to the top and **drop** it into the desired ordered position.

3. Please see the initial segments on the right panel. These segments have been found to occur frequently in the given dataset.

4. You can define initial entries for each of your attributes. You can do this by **dragging** entries from the frequent segments (in the right panel), and **dropping** them to the appropriate attribute box.

Candidate segments for initializing the constraints:

wxp pro
ddk qbe
rdd notebook d.ddghz
ddgb dd" cd rw
xpp dd xga
d.d ghz
integrated intel
x p mmx
notebook d.ddghz
xp pro
no rtns
dd" xqa

[Add an attribute] [Remove selected]

Attribute name	Initial entries, separated by comma
☐ Models	xdd, tdd, rdd, notebook, i series
☐ Processors	piii, intel pentium iii,
☐ CPU Clock Speed	d.dghz, dddmhz

[Create Constraints]

Fig. 15. Screen shot from our web based dictionary discovery tool.

Attribute dictionaries help to identify structure in un-structured and semi-structured records, by providing a list of valid values for a specific domain, and filtering values that may be irrelevant to this domain. They also help to enforce a consistent representation of the domain values by having a single source where all the valid values of an attribute are specified. These benefits make attribute dictionaries extremely useful towards improving product search, and for improving and maintaining data quality.

We envision three lines of research as next steps. First, we currently assume that the attribute domains are non-overlapping. It would be interesting to evaluate our techniques on data sets that have overlapping domains and consider extensions to the framework that take into account overlap in domains. Second, we can consider methods for automatic extraction of the initial dictionary seed values from online text sources and knowledge bases where techniques (such as Hearst patterns) may be used to identify different types of semantic relationships for a given domain. Finally, in our dictionary population step, we intend to take into account the semantic similarity of tokens already assigned to a dictionary. A dictionary may, for example, contain values that are all nouns representing cities. Using semantic similarity, we can reject tokens that are found to belong in the dictionary but are of different type. In this phase, we may similarly use other repeating patters, as well as part-of-speech tags.

References

1. Yelp dataset challenge (2011). www.yelp.ca/dataset_challenge
2. General electric second annual major purchase shopper study. GE Capital Retail (2013)
3. Agathangelou, P., Katakis, I., Kokkoras, F., Ntonas, K.: Mining domain-specific dictionaries of opinion words. In: Web Information Systems Engineering, pp. 47–62 (2014)
4. Agrawal, R., Imielinski, T., Swami, A.N.: Mining association rules between sets of items in large databases. In: SIGMOD Conference, pp. 207–216 (1993)
5. Bing, L., Lam, W., Wong, T.-L.: Wikipedia entity expansion and attribute extraction from the web using semi-supervised learning. In: International Conference on Web Search and Data Mining, WSDM, pp. 567–576 (2013)
6. Borkar, V., Deshmukh, K., Sarawagi, S.: Automatic segmentation of text into structured records. SIGMOD Rec. **30**(2), 175–186 (2001)
7. Chaturvedi, S., Prasad, K.H., Faruquie, T.A., Chawda, B., Subramaniam, L.V., Krishnapuram, R.: Automating pattern discovery for rule based data standardization systems. In: ICDE, pp. 1231–1241 (2013)
8. Chiang, F., Andritsos, P., Zhu, E., Miller, R.J.: Autodict: automated dictionary discovery. In: ICDE, pp. 1277–1280 (2012)
9. Cohen, W.W., Sarawagi, S.: Exploiting dictionaries in named entity extraction: combining semi-markov extraction processes and data integration methods. In: SIGKDD, pp. 89–98 (2004)
10. Cortez, E., da Silva, A.S., Gonçalves, M.A., de Moura, E.S.: Ondux: on-demand unsupervised learning for information extraction. In: SIGMOD Conference, pp. 807–818 (2010)
11. Cover, T., Thomas, J.: Elements of Information Theory. Wiley, New York (1991)
12. Godbole, S., Bhattacharya, I., Gupta, A., Verma, A.: Building re-usable dictionary repositories for real-world text mining. In: CIKM, pp. 1189–1198 (2010)
13. http://crf.sourceforge.net/
14. Klein, D., Manning, C.: Accurate unlexicalized parsing. In: Proceedings of ACL, pp. 423–430 (2003)
15. Lafferty, J.D., McCallum, A., Pereira, F.C.N.: Conditional random fields: probabilistic models for segmenting and labeling sequence data. In: ICML, pp. 282–289 (2001)
16. Lee, T., Wang, Z., Wang, H., Hwang, S.-W.: Attribute extraction and scoring: a probabilistic approach. In: ICDE, pp. 194–205 (2013)
17. Li, G., Deng, D., Feng, J.: Faerie: efficient filtering algorithms for approximate dictionary-based entity extraction. SIGMOD 2011, pp. 529–540 (2011)
18. Li, X., Wang, Y.-Y., Acero, A.: Extracting structured information from user queries with semi-supervised conditional random fields. In: SIGIR 2009, pp. 572–579 (2009)
19. Nguyen, H., Fuxman, A., Paparizos, S., Freire, J., Agrawal, R.: Synthesizing products for online catalogs. Proc. VLDB Endow. **4**(7), 409–418 (2011)
20. Pantel, P., Philpot, A., Hovy, E.H.: An information theoretic model for database alignment. In: SSDBM, pp. 14–23 (2005)
21. Peshkin, L., Pfeffer, A.: Bayesian information extraction network. In: IJCAI 2003, pp. 421–426 (2003)
22. Rissanen, J.: Modeling shortest data description. In: Automatica (1978)
23. Roy, S., Chiticariu, L., Feldman, V., Reiss, F., Zhu, H.: Provenance-based dictionary refinement in information extraction. In: SIGMOD Conference, pp. 457–468 (2013)

24. Sarawagi, S., Cohen, W.W.: Semi-markov conditional random fields for information extraction. In: NIPS, pp. 1185–1192 (2004)
25. Sarkas, N., Paparizos, S., Tsaparas, P.: Structured annotations of web queries. In: SIGMOD Conference, pp. 771–782 (2010)
26. Slonim, N., Tishby, N.: Agglomerative information bottleneck. In: NIPS, pp. 617–623 (1999)
27. Socher, R., Bauer, J., Manning, C., Ng, A.: Parsing with compositional vector grammars. In: Proceedings of ACL, pp. 455–465 (2013)
28. Sutton, C., Mccallum, A.: Introduction to Conditional Random Fields for Relational Learning. MIT Press, Cambridge (2006)
29. Tan, B., Peng, F.: Unsupervised query segmentation using generative language models and wikipedia. In: WWW, pp. 347–356 (2008)
30. Winkler, W.E.: String comparator metrics and enhanced decision rules in the fellegi-sunter model of record linkage. In: Survey Research, pp. 354–359 (1990)
31. Zhang, M., Hadjieleftheriou, M., Ooi, B.C., Procopiuc, C.M., Srivastava, D.: Automatic discovery of attributes in relational databases. In: SIGMOD Conference, pp. 109–120 (2011)
32. Zhang, Z., Zhu, K.Q., Wang, H., Li, H.: Automatic extraction of top-k lists from the web. In: ICDE, pp. 1057–1068 (2013)

Subject-Related Message Filtering in Social Media Through Context-Enriched Language Models

Alexandre Davis[(✉)] and Adriano Veloso

Universidade Federal de Minas Gerais, Avenida Presidente Antônio Carlos,
6627, Belo Horizonte, MG 31270-901, Brazil
{agdavis,adrianov}@dcc.ufmg.br
http://www.ufmg.br

Abstract. Efficiently retrieving and understanding messages from social media is challenging, considering that shorter messages are strongly dependent on context. Assuming that their audience is aware of background and real world events, users can shorten their messages without compromising communication. However, traditional data mining algorithms do not account for contextual information. We argue that exploiting context can lead to advancements in the analysis of social media messages. Recall rate increases if context is taken into account, leading to context-aware methods for filtering messages without resorting only to keywords. A novel approach for subject classification of social media messages, using computational linguistics techniques, is proposed, employing both textual and extra-textual (or contextual) information. Experimental analysis over sports-related messages indicates over 50 % improvement in retrieval rate over text-based approaches due to the use of contextual information.

Keywords: Computational linguistics · Information retrieval · Social media

1 Introduction

In a recent phenomenon, many people have started using online social media as their primary channel for interaction and communication. Social interaction with other people has been partly transferred to online social networks, in which communication can take place between a single sender and large numbers of receivers, using text messages along with images and video. The use of mobile devices as instruments for this kind of communication allowed the subject of such messages to be associated to events and phenomena of immediate (and usually ephemeral) interest, potentially covering all kinds of themes and topics. There are billions of users of social media services, and the number of messages grows continuously. In only one of the most popular social networks, the number

© Springer-Verlag Berlin Heidelberg 2016
N.T. Nguyen et al. (Eds.): TCCI XXI, LNCS 9630, pp. 97–138, 2016.
DOI: 10.1007/978-3-662-49521-6_5

of daily exchanged messages reached 64 billion in late 2014[1]. In another, the number of active users exceeds 1.3 billion (one of every 5 humans), and over 300 million images are uploaded and distributed every day[2].

Since in this kind of communication messages can be digitally encoded, stored, accumulated and analyzed as a stream of data, many interesting research opportunities arise, associated to the challenge of handling large volumes of real-time data and fueled by the overall goal of increasing our understanding of human societies. Suppose that a social scientist becomes interested in analyzing the repercussion of some theme, or subject, of current interest. If it is possible to, routinely, (1) tap into a social media source, (2) filter out messages that are associated to the selected subject, and (3) analyze the set of messages according to parameters such as content, vocabulary, volume, emotional bias, propagation speed, spatial distribution, duration, and others, then the scientist can frequently have instantaneous glimpses on the way society treats that subject and can advance current knowledge for similar situations in the future. Using large volumes of data from social networks, social sciences achieve a scale that has never been possible before, in one of the many facets of the so called *data science* phenomenon [1].

However, there are numerous difficulties in every step of this process. In the first step, not every social media provider allows broad access to messages, user profiles and user relationships. The second and third steps introduce complex challenges as to accurately selecting messages as being related or not to a given subject, and on analyzing sets of messages related to a given theme. Furthermore, the sheer volumes of messages that are exchanged daily indicate that only automated approaches can be used in a meaningful way, and that the complexity of these approaches is limited by the computational infrastructure that is made available to run them.

Consider, specifically, the case of Twitter. It is the fifth social network as to the number of users, but its characteristics make it a primary source for research initiatives. Twitter offers a comprehensive Application Programming Interface (API), with clear rules for collecting and filtering messages in significant volumes. User profiles can be obtained, and user relationships can be retrieved. On the other hand, messages are limited to 140 characters, and thus issues such as abbreviations, misspelled words and informal lexical and grammatical constructions are common. Twitter users also frequently rely on contextual information[3] as a way to keep messages short. Assuming that the receivers will be able to recognize the context, users frequently omit key parts of the message, posting incomplete sentences or chaining ideas over several messages spread through time.

Regardless of such difficulties, filtering and analyzing short messages in a medium such as Twitter has been successfully done in the recent past, with

---

[1] WhatsApp, according to http://www.statista.com/statistics/258743/daily-mobile-message-volume-of-whatsapp-messenger/.

[2] Source: https://zephoria.com/social-media/top-15-valuable-facebook-statistics/.

[3] By *context* we mean all non-written information that is relevant to understand a message, such as real world events and common sense knowledge.

significant results regarding problems such as sentiment analysis [2], spatial location [3] and epidemics forecasting [4] (all of these are results from our research group at UFMG). A broad range of techniques has been used in those studies, including data mining, machine learning, and natural language processing (NLP). In most of these cases, filtering has been simplified to use sets of keywords, in a strategy applied in large scale. Contextual information has usually not been taken into consideration. In part, the success of such initiatives comes from the fact that the number of messages for which context is not necessary is still large, allowing for the recognition of trends and for the identification of relevant information.

Consider, for example, an individual social media user A, that at some point in time issues a message containing the sentence "And he scores!!". Without any contextual information, an automated filtering or analysis agent would have trouble associating such a message to any possible subject. However, the receivers of the message, individuals B_i in A' s social network, know from previous interactions that A is a soccer fan, who roots for a specific team T. Some of those may further know that a game is going on at the moment the message is issued, therefore deducing that some player of T has scored. Therefore, a message containing only three words and only 15 characters can convey a lot of information to its receivers, but most of that information has to be gathered from the context, including information on user previous behavior, preferences, habits, and external information on game schedule, news and much more. A keyword-based filtering system would have missed all this information, losing a valid indication of a crucial moment (scoring) in the event (the soccer match)[4]. Humans can do that in a natural way, but transferring this kind of possibility to a computer represents a big challenge.

For this work, therefore, we pose some questions: how relevant is contextual information in the successful filtering of messages that are relevant to a selected subject? Can the recall rate increase if context is taken into account? Furthermore, if the context is known and can be detected, is it possible to filter relevant messages without resorting only to keywords? In this paper, we propose a novel approach for subject classification of social media messages that considers both textual and extra-textual (or contextual) information. Techniques based on concepts of computational linguistics, more specifically in the field of *pragmatics*, are employed. For experimentally analyzing the impact of the proposed approach, different datasets containing messages about three major American sports leagues (football, baseball and basketball) were used.

1.1   Statement

Based on Pragmatics theory, we posit that text is not informative enough for subject classification in social media. Therefore, we need extra-textual informa-

---

[4] More than 35 million tweets were issued during the Brazil vs Germany match in the FIFA World Cup 2014 semifinals, an event of global repercussion, the most tweeted event on record so far – but who knows how many tweets are not included in this figure due to lack of contextual information?.

tion to supplement written message in this scenario to achieve better classification recall. In this paper, we propose contextual models (from Pragmatics) and language models that approximate human reasoning to include contextual information in message interpretation. Finally, we claim that messages uttered in personal and informal media require this kind of information to be fully understood.

1.2   Objectives

The main objective of this work is to demonstrate the importance of context to infer the subject of social media messages. We propose a novel method for identifying subjects, considering both textual and extra-textual information. This method is an adapted language model that uses context models (also proposed as part of this work) to supplement written information.

With the proposed techniques, our goal is to show that it is possible to increase the recall of subject-related messages in social media without using a bag-of-words data gathering model. We posit that retrieval based on fixed keyword selection is misleading for some kinds of social media messages, specifically the ones that are deeply affected by pragmatic effects and *conversational implicatures*, i.e., elements of language that contribute to understanding a sentence without being part of what is said, nor deriving logically from what is said.

1.3   Specific Objectives

The following specific objectives were pursued in this work:

- To study and organize concepts related to computational linguistics, more specifically about pragmatics, in order to apply them to the problem of filtering social media messages;
- To obtain, organize and annotate a corpus of messages pertaining to popular sports (football, baseball and basketball), with which to analyze and measure the effect of context;
- To conceive and implement filtering algorithms that are able to consider contextual information;
- To experimentally analyze the proposed techniques, quantifying their effect over the test data, and obtaining valuable insights as to the application of the proposed techniques in other applications.

1.4   Challenges and Contributions

The challenges and contributions of this work follow:

- **Data gathering without keywords:** Retrieving a tweet stream without using bag-of-words selection was a great challenge. Twitter's API only provides three options: (1) retrieve all messages that contain a set of keywords, (2) retrieve 1 % of the entire message stream and (3) retrieve all messages posted by a set of users. Since we want to identify subjects in tweets without resorting to keywords, and in the second option we would get an extremely sparse stream, the only option left was to select by users. We proposed a method and demonstrated some hypotheses for choosing a particular set of users and enabling the use of this stream.

- **Contextual elements in the labeling process:** It is important, by our hypothesis, to consider contextual elements for labeling test messages. Since human annotators are not on the same place, time and may not have the same knowledge as the original receivers of the Tweet, it may get a different interpretation of the message. Therefore, we tried to reproduce as much as possible the original post's context in our label environment by displaying user profile (e.g. avatar photo, background image), previous posts, and considering the time of posts to the human rater to give a better background understanding. This is an uncommon approach in the literature, which usually restricts rating to text only.
- **Pragmatic contextual models:** Contextual models are an innovative way of scoring the likelihood of a message to be related to a subject according to its contextual information. The score generated by these models can be used alone or associated with a language model.
- **Context-enriched language models:** To combine the novel pragmatic contextual model with the text posted in messages, we propose a new language model. The idea of combining non-textual elements with text is a major contribution of this work.

1.5   Structure

The remainder of this paper is organized as follows. Section 2 covers literature that is relevant to our proposal and explores some basic concepts. Sect. 3 presents a discussion on pragmatic effects in social media messages, and introduces methods for modeling context. Next, a new language model that uses these contextual scores is proposed in Sect. 4. Sections 3 and 4 contain the major contributions of this paper. Section 5 contains a description of the datasets used in experimentation, along with demonstrations of the proposed contextual models and empirical demonstrations on the validity of assumptions made in Sect. 3. Section 6 presents the results of the experimental evaluation of the proposed language model. Finally, Sect. 7 shows conclusions and discusses future work.

2   Related Work

In this section, we present some literature related to this work. We start with a section discussing pragmatic effects and conversational implicatures, which are our main motivation to claim the importance of context for social media messages interpretation. Then, we discuss language models, which are the foundations of the proposed technique. Next, we show other works on text categorization, a more general version of the subject classification problem. Finally, we show some initiatives that tried to include information from different sources to enrich text when it is scarce.

2.1   Pragmatic Effects and Conversational Implicatures

Contextual influence over text has been studied by linguists, comprising the so-called *pragmatic effects* [5]. Pragmatic effects are related to the human ability to use context for changing the uttered message's semantic meaning. They are frequently present in communication and may manifest themselves in several different ways, according to the medium (e.g. written, oral, computer mediated), subjects involved, situation of communication, previous shared knowledge between subjects, and others. Pragmatics effects have a major role in short and informal messages, such as those that are common in social media.

Pragmatics effects may manifest in many different sources of information. In some situations, pragmatics effects are used as a form of generating humor [6]. In this case, the change in the sentence meaning generates an unexpected or ironical interpretation. Another interesting pragmatic effect derives from facial expressions and eyeball movements. Humans are able to detect minimal movements in eyeball that indicate changes in the meaning of uttered messages [7]. However, one of the most common pragmatic effects are the *conversational implicatures* [8,9].

Conversational implicatures are an implicit speech act. In other words, they are part of what is meant by a speaker's utterance[5] without being part of what was uttered. Implicatures have a major contribution to sentence meaning without being strictly part of 'what is said' in the act of their enunciation, nor following logically from what was said. Therefore, implicatures can be seen as a method of reducing the gap between what is literally expressed and the intended message [5]. A good example of implicatures is given by Levinson [5]. If a speaker A tells receiver B "The cat is on the mat", without any contextual information, this sentence may be interpreted as a simple statement and no action is required. However, if both A and B know that this cat usually goes to the mat when it is hungry, the uttered message may be interpreted by B as "The cat is hungry. Please feed it". This example is interesting because the intended meaning of the message is completely modified once both A and B share a common piece of information.

Conversational implicatures are essential to expand oral language expressiveness, since humans have a tendency to contract spoken messages as much as possible without compromising message comprehension [5,9]. Conversational implicatures, however, are not exclusively used in spoken communication, they are also important on computer mediated communication (CMC), especially on social media (such as blogs, instant messaging, and in social media messages typical of Twitter and Facebook, among others) [10]. As previously mentioned, in this kind of media users want to post short and concise messages because of time and space constraints.

Despite of being really important for informal and personal communication, most of the natural language processing (NLP) approaches overlook conversational implicatures. In this work, we propose to consider implicatures in language models, one of the most traditional NLP techniques. In the next section, we discuss some previous works on language models.

---

[5] An uninterrupted chain of spoken or written language.

2.2 Language Models

Statistical language models are an important class of algorithms in natural language processing (NLP) literature. The main goal of these models is to estimate the probability of finding a given sequence of words in a set of sentences, commonly called in this context a *language*. One of the most traditional methods for estimating that probability is the n-gram language model [11]. This model assumes that words in the input sequence are not independent. Moreover, it argues that the probability of a word w in the input sequence depends only on its textual context C (usually described as the n words preceding w). Therefore, the n-gram language model estimates the probability $P(w|C)$, which is useful in many applications, such as automatic translation [11] and speech recognition [12]. For these applications, however, methods that consider $P(w|C) \neq P(w)$ perform much better than those that assume independence between words in a sentence [13]. We argue that, in social media text, conversational implicatures make textual context C unreliable to estimate this probability. For this matter, we propose to add extra-textual context to improve the accuracy of language models for this specific kind of data.

Another interesting aspect of language models is that they can be used as an unsupervised learning method, since they use a set of unlabeled sentences in their construction. This characteristic makes them useful for information retrieval and text categorization applications. Many information retrieval techniques, for instance, improve the ranking of Web pages or texts by using language models [14]. In these cases, the input word sequence is given by the user while constructing a query. Regarding text categorization, some authors proposed approaches that use language models for improving robustness in texts with misspelled words and syntactic errors [15], such as those extracted from e-mails and from documents generated by optical character recognition (OCR). Alternatively, language models may be used for clustering similar documents into categories [16] and for improving hypotheses in vastly used classifiers, such as Naive-Bayes [17] and Logistic Regression [13]. In the next section, we show some other approaches for text categorization.

2.3 Text Categorization

Text categorization (or classification) is a classic problem in Machine Learning and NLP. The goal is to assign previously defined labels, or classes, to fragments of real-world texts. Depending on the applications proposed, those labels may also be called topics. In early works, news corpora and medical records have been used as textual data sources [18,19]. Recently, text categorization approaches have been adapted to challenging datasets, such as blog texts [20,21] and spam filtering [22,23]. Some challenges associated with these new datasets are text sparsity and oralism, which lead to adaptations in traditional machine learning algorithms.

Traditional machine learning text classification approaches have addressed some of these issues by using a bag-of-words model, in which each word is used as a feature for a supervised classifier [24]. Unfortunately, this model produces an enormous number of dimensions, which may impact in the classifier's accuracy [25]. Therefore, authors have proposed dimensionality reduction techniques [19,26] for achieving better results. Other authors argue that generative models are more robust to unstructured text and to situations where a document may have multiple labels [27,28]. Such approaches are heavily dependent on textual features and on big training sets. Thus, it would be unfeasible to use them on social media message streams. Our proposal uses contextual features and unsupervised methods (language models). Consequently, our approach is less dependent on text quality and does not require any previous manual labeling effort.

Recently, incremental online classifiers were proposed for solving text categorization in scenarios that require constant updates in training data [29,30]. Another possibility is to automatically update online classifiers with techniques such as EM^2 [31]. However, in some datasets, such as blogs and social media text, changing the training set may not be enough. Many authors addressed the vocabulary dynamicity and text sparsity problems using semantic analysis [32–34]. We believe that this approach is not effective in social media datasets, given the number of grammatical and spelling errors usually found in this kind of text. Therefore, in these cases, we may need to find other information sources to complement or substitute textual information. In the next section, we show some works that used alternative data sources for improving the accuracy of categorization and classification systems for scenarios in which the textual information is scarce or incomplete.

2.4   Alternative Data Sources

In applications for which textual data is sparse, some works resort to external sources (including knowledge bases such as Freebase, WordNet and Wikipedia) to help in the categorization process [35–38]. Some authors also use custom ontologies as input for their methods [39]. Another usage for external knowledge sources is to relate entities using inference rules [36,40]. Although this is a good way of contextualizing information, we observe that external sources are not updated at the same pace as new expressions and terms are created in social media (for instance, slang, nicknames, ironic expressions). Alternatively, we may search for contextual information within the dataset itself [41]. One contribution of our proposed approach is to extract this information using mostly meta-information on the messages. Our goal is to use this kind of information to mimic conversational implicatures, as described by Pragmatics Theory [9], commonly used in human communication.

According to Pragmatics Theory, as texts become shorter and more informal, environment knowledge becomes more important to discover the message's meaning [42]. Moreover, in instant messaging and online chat discourse, users have a tendency to contract the text as much as possible with the objective of typing faster or mimicking spoken language [43]. For instance, character repetition is a very common way of reproducing excitement in this kind of text (e.g.

"Goooooooal") [44]. In social media, the sender's knowledge of real-world events [45], location [38] and popularity [46] may influence the elaboration of messages, assuming a similar set of knowledge on the part of the receiver. Bayesian logic has been used to extract "common-sense" beliefs from natural language text [40]. A recent work has proposed combining implicit information such as tone of voice and pauses in sentences for improving the results in speech translation techniques [12]. Our proposal is to model this implicit external influence in social media messages, and to use it to classify each one according to predefined labels, or subjects.

3 Pragmatic Contextual Modeling

Previous pragmatics studies argue that context is an important information source for understanding the message's utterance. One of the most common examples is spoken communication, where, the meaning of the speaker's message is frequently not clear in the utterance and, therefore, requires additional knowledge to be interpreted. For better comprehending the dimensions of communication, Grice introduced the *cooperative principle* [47], a major contribution to pragmatics theory, which describes how people communicate with each other. This principle defines assumptions or maxims on the nature of the information that is expected to be exchanged in conversation (Definition 1).

Definition 1 (Cooperative Principle). *Pragmatic contributions to sentence meaning should follow the accepted purpose or direction of the talk exchange in which the speaker is engaged. Therefore, contributions should usually respect the following maxims:*

Maxim of Quantity: *The author provides only the amount of explicit information necessary for the listeners to complete the communication.*
Maxim of Quality: *There is no false information.*
Maxim of Relevance: *The information needs to be relevant to the conversation topic.*
Maxim of Manner: *The information should be clear.*

Another important contribution to pragmatics is Hirschberg's definition of *proposition* in conversational implicatures [48]. According to Hirschberg, implicatures are composed by propositions, which are part of the inference process that the receiver is required to do if he believes the sender is following the cooperative principle. Therefore, propositions can be seen as bits of information that are considered by the sender to be known by the receiver as well. If this assumption is false, the cooperative principle fails and, usually, there is a communication break. In some cases, however, violations in the cooperative principle are used for humor and irony [49,50]. Because of that, violation in Maxim of Quality is common in social media.

Definition 2 (Hirschberg's Proposition). *Given a utterance U and a context C, proposition q is part of a conversational implicature of U by agent B if and only if:*

1. *B believes that it is public knowledge in C for all the discourse participants that B is obeying the cooperative principle.*
2. *B believes that, to maintain item 1 given U, the receiver will assume that B believes q.*
3. *B believes that it is mutual and public knowledge for all the discourse participants that, to preserve item 1, one must assume that B believes q.*

Definitions 1 and 2 can be better explained using an example, as follows. During a football match someone listens agent B uttering *"Oh NO!"* (this is U). Unconsciously, the recipient of U infers that *"Something bad happened with the team B cheers for"* (this is q). The context C, in this example, informs not only that there is a match happening at utterance time, but also the team B cheers for. Therefore, q follows Definition 2.(1), by which B expects that the discourse participants are aware of the necessary knowledge in C to understand the message.

Both the cooperative principle and Hirschberg's definition guide some properties in conversational implicatures that are essential to the proposed contextual models. The first important property is that implicatures are **non-lexical**; therefore, they cannot be linked to any specific lexical items (e.g. words, expressions). Second, implicatures **cannot be detached from utterances** by simple word substitution. In the football match example, it is clear that implicatures are not linked to any word in U (i.e., the proposition would be the same if the utterance was different, but issued to the same effect). Another important property is **calculability**: the recipient should be able to infer implicatures from utterances. In the football example, the recipient of the message needs to be able to deduce to which team something bad is happening.

Conversational implicatures and the cooperative principle, however, are not exclusively used in oral communication. It is clear that both of these theories apply to computer mediated communication (CMC), especially in social media [42]. Twitter messages provide many clear situations that fit these theories. For instance, the maxim of quantity is usually present in a considerable share of the messages, due to message size limitations - messages must have less than 140 characters. Also, the maxim of relevance is essential in Twitter, since messages are closely related to real time events. Finally, it is important to notice that the sources of contextual information in social media are not as rich as in spoken communication. Consequently, users add new forms of context to increase their expressiveness in this media, such as emoticons, memes, hashtags, videos and photos. All these characteristics make it even harder to understand the meaning of a social media message just by reading the plain text.

Recently, we have seen many approaches for sentiment analysis [51] and topic tracking [52] on social media. Unfortunately, most of these proposals completely ignore extra-textual features. Many of those authors, for instance, use as data source a stream of messages selected by keywords. Under the conversational implicature hypothesis, we argue that keyword selection is not enough for filtering a subject-related message stream. Therefore, many of these works are using biased data streams with limited recall. In Sect. 5 (Table 3), we show an estimate of how much data is lost using keyword-selection. Our objective is to increase recall in Twitter subject-related data gathering. This approach relies on contextual scores given to each tweet. These contextual scores were proposed to imitate the role of information that is commonly used in conversational implicatures.

In this work, we focus on two major types of contextual information for social media communication: *agent knowledge context* and *situational context*. Agent knowledge context compresses information about an agent that can be accessed by all discourse participants (according to Definition 1). For this category, it is expected that we have information such as agent interests in subjects (e.g. american football, baseball, basketball). On the other hand, situational context compresses information about real-world events that are seen as common knowledge between all discourse participants, such as football matches, scoring situations (e.g. touchdowns and goals), team news (e.g. player hiring and injury reports). In the next sections, we propose models to evaluate the relation of each of these types of context to a subject in a Twitter message. These models generate a numeric score that indicates the model's confidence on the relatedness between subject and message context.

3.1   Agent Knowledge Context

A user in social media is generally interested in several subjects. It is straightforward to assume that a user interested in american football will issue messages about this subject many times. Therefore, receivers would often be able to identify the interests of a given user by looking at the distribution of messages related to each subject over time. It is also interesting to notice that knowledge about a user is built based on a longer term relation, needed to increase the quality of the contextual information the receivers have about the user.

The proposed agent knowledge model tracks the amount of messages uttered by a user b in a long period training stream related to the subject S. As argued throughout this paper, due to some characteristics in communication, it is hard to identify whether a message is related to S. For that matter, we use an heuristic to compute the proposed agent knowledge model that is based on the frequency of manually selected keywords K_S related to S in previous messages from b. We select keywords K_S that are trivially related to the subject S, the criteria will be better explained in Sect. 5 (The exact selected keywords can be further analyzed in Attachment A). This heuristic considers two important hypotheses, as follows:

Hypothesis 1. The chance of a user b being interested in a subject S is proportional to the frequency of keywords from K_S that have been used by b related to S.

Hypothesis 2. Ambiguous keywords are used by more users (including users that are not interested in S) than unambiguous ones.

The intuition behind Hypothesis 1 is that if a user b frequently utilizes words in K_S, it is probable that b is interested in S. For instance, a user that, over two months, has posted a hundred keywords related to baseball is more likely to be interested in baseball than a user that posted only ten keywords of the same set in that period. However, these keywords can be ambiguous, (i.e., keywords in K_S can also belong to K_t, where T is a subject that is different from S) and Hypothesis 2 tries to neutralize this.

For understanding the intuition behind Hypothesis 2, consider two sets of users: U_S are the users interested in S and $\overline{U_S}$ are everyone else. It is expected that more ambiguous keywords (such as "NY", "Boston") are used by both U_S and $\overline{U_S}$, while unambiguous keywords (such as "Packers", "Red Sox") are referenced mostly by users in U_S. Since $|U_S| << |\overline{U_S}|$, ambiguous keywords are referenced at least once by a wider range of users.

With Hypotheses 1 and 2, we can propose a score for the relation between an agent and a subject S given the agent knowledge context. The score is similar to the traditional TF-IDF (term frequency - inverse document frequency) measures used in information retrieval. Following Hypothesis 1, we define a value TF_k^b (Eq. 2) that represents the normalized number of times that a keyword $k \in K_S$ was used by a user b in the analyzed period. For each keyword, we define a value IDF_k (Eq. 1), which gives a weight to k according to the ambiguity Hypothesis 2. These values are calculated using the following formulas, where N is the number of users, n_k is the number of users that used k at least once in the given stream and f_k^b is the number of times k was uttered by user b.

$$IDF_k = log(\frac{N}{n_k}) \tag{1}$$

$$TF_k^b = 1 + log(f_k^b) \tag{2}$$

Finally, we define a score AKC^b for each user b in the stream, defined by the product of IDF_k and TF_k^b (Eq. 3). Information retrieval approaches normalize the TF-IDF score by the document size. As described ahead, in Sect. 5, the stream used for model generation (i.e. training stream) was created by keyword projection. Therefore, it is impossible to normalize this value by the number of messages posted by b, since the dataset contains only messages that include at least one keyword $k \in K$.

$$AKC^b = \sum_{}^{\forall k \in K_S} (IDF_k * TF_k^b) \tag{3}$$

Having thus defined the agent knowledge context, which will allow us to quantify the effect of the user's knowledge of a subject, next section approaches the situational context, by which the effects of concurrent events on the utterance of messages will be evaluated.

3.2 Situational Context

Social media messages are often closely related to real world events. We argue that most users expect that their receivers are aware that a real world event at the time of an utterance may be an important contextual information for the posted message. Consequently, we posit that tweets posted during important events of S are more likely to be related to S (Hypothesis 3).

Hypothesis 3. Messages are more likely to be related to a subject S if they were posted within a timeframe that contains meaningful real world events related to S.

In the proposed situational model, our goal is to measure the relation of a message posted during a time window T with subject S^6. Given Hypothesis 3, we want to assign higher scores for messages that were posted during important events related to S. Unfortunately, since most subject-related events are highly dynamic, it is unfeasible to extract such contextual knowledge from external sources. In this section, we define methods for estimating the likelihood of such events to be happening during T given the messages posted in that period of time.

For estimating this likelihood, we hypothesize that events are more likely to occur in timeframes in which a higher frequency of unambiguous keywords related to S is observed (Hypothesis 4). For instance, during American football season matches, a much higher frequency of unambiguous football-related keywords are expected, such as "Green Bay Packers", "GoPats" and "New York Jets". If there was no match happening, the frequency of those keywords would be much lower. It is important to notice that matches are not the only important events in this scenario, breaking news, hiring and draft may also impact the Situational Context of a message.

Hypothesis 4. The probability of a meaningful event related to S to be occurring during time window T is correlated with the number of unambiguous keywords related to S that are posted during T.

Following Hypothesis 4, we define a score SC^T (Eq. 5) for each fixed time window T, to measure the likelihood of a meaningful event related to S to be happening during T. To calculate this score, we use TF_k^T to track the frequency of a keyword k during a time window of T. Finally, we normalize the score with the number of tweets in the training stream that have been issued during that time window.

$$TF_k^T = 1 + log(f_k^T) \tag{4}$$

$$SC^T = \sum^{\forall k \in K_S} \frac{IDF_k * TF_k^T}{|T|} \tag{5}$$

It is important to notice that, despite of being TF-IDF-based, situational context and agent knowledge are not correlated. Therefore, these contextual sources may be used simultaneously by human cognition in the message interpretation process. Consequently, to improve even further the ability to reproduce cognition using pragmatics, we need methods for combining contextual features.

---

[6] For simplicity, fixed-length time windows are adopted.

Analyzing the potential in act of mixing multiple contextual features is left for future work. In the next section, we propose a novel language model that uses these proposed scores for improving retrieval performance.

4   Language Model

As discussed previously, conversational implicatures are a common pragmatic effect in social media text. Space constraints and speech informality increase the importance of context for interpreting short messages. Unfortunately, most of the traditional NLP and Machine Learning techniques ignore extra-textual information, mainly because they were developed for self-contained textual sources, such as news and books. One of the most important classic NLP techniques, Language Models have been used as abstract methods of representing language patterns. This representation involves using probability estimations for the presence of a word within a sentence. We propose a novel approach for estimating such probabilities using extra-textual information. This approach is detailed in Sect. 4.2, after reviewing the classic n-gram language model.

4.1 n-gram Language Model

The traditional n-gram language model assumes that the probability of a word w to be part of a language depends only on the previous $n - 1$ words. These sequences of words are called n-grams. This way, the goal of a n-gram language model is to estimate the probability $\hat{P}(w_i|w_{i-1}, w_{i-2}, ..., w_{i-n})$, given L, a training set of text (commonly called a *language sample*). The probability of a text fragment, or message, m to be part of a language is given by the following formula:

$$P(m \in L) = \prod^{w_i \in m} P(w_i|w_{i-1}, ..., w_{i-n}) \tag{6}$$

A major concern about this model is that if a n-gram is not found in the training examples, the maximum likelihood estimator would attribute a null probability to this n-gram and, consequently the whole text fragment would get no probability of belonging to the language. Many previous works have used smoothing methods to force $P(w_i|w_{i-1}, ..., w_{i-n}) > 0$. For this work, we are using Katz backoff with Lidstone Smoothing, described in the next subsections, to estimate the probability of a n-gram belonging to the language.

Lidstone Smoothing. Lidstone smoothing (or additive smoothing) is a technique used by the probability estimator to attribute a probability for items unseen in the training set. The main idea of this method is to add a constant value α for each event in the amostral space. This way, every time the estimator receives an n-gram not found in the training examples, it attributes a very small but greater than zero probability to the n-gram.

Consider c_i as the number of times an event i happens, and C as the amostral space ($\sum c_i = N, c_i \in C$). The estimated probability for the event i is given by a small modification to the Maximum Likelihood Estimator (MLE) (Eq. 7). Notice that for each event in the amostral space, a value α is added independently from the event's frequency. Therefore, if $c_j = 0$, $\hat{P}(j) = \alpha/(N + \alpha|C|)$.

$$\hat{P}(i) = \frac{c_i + \alpha}{N + \alpha|C|} \qquad (7)$$

The choice of α is essential for the smoothing quality, since bigger values imply in attributing a larger share of the probability space to unseen examples. Consequently, if α is too large, events with low frequency will be attributed a probability that can be similar to those with high frequency. This is called *smoothing underfitting*. In this paper, we chose the best α experimentally.

For the language model, we need to adapt this estimator to be used with conditional probabilities. To estimate the probability $\hat{P}(w_i|w_{i-1}, ..., w_{i-n})$, we also adapt from MLE. Therefore, we have the estimated probability in Eq. 8, where $f_{w_i,...,w_{i-n}}$ is the frequency of a n-gram and $f_{w_{i-1},...,w_{i-n}}$ is the frequency of the $(n-1)$-gram in conditional probability prior.

$$\hat{P}(w_i|w_{i-1}, ..., w_{i-n}) = \frac{f_{w_i,...,w_{i-n}} + \alpha}{f_{w_{i-1},...,w_{i-n}} + \alpha|C|} \qquad (8)$$

Katz Backoff. When a n-gram cannot be found within a language model, it is interesting to notice that we may generalize it into a simpler $(n-1)$-gram, which is more likely to have occurred at least once in the dataset. The generalization property is an important feature of this kind of language model, enabling recursive search for a textual token in simpler models, also known as *backoff*. With the backoff, n-gram based language models are able to attribute a probability for a not found item that is proportional to the probability of finding it in a more general model. This is expected to perform better than attributing a constant probability, as in Lidstone smoothing. For this work, we use Katz backoff, which is a method for estimating the probabilities $\hat{P}(w_i|w_{i-1}, ..., w_{i-n})$.

Katz backoff defines N levels of language models according to the generality. The first level is the most specific language model, which estimates $\hat{P}(w_i|w_{i-1}, ..., w_{i-n})$. If $f_{w_i} = 0$ or $f_{i-1,...,i-n} = 0$, backoff to the next level is needed. Similarly, the next level estimates $\hat{P}(w_i|w_{i-1,...,i-n-1})$. In the last level, we get the most general language model that estimates the probability of a unigram (or single token) belonging to the language model: $\hat{P}(w_i)$. The discount and backoff processes are illustrated in Fig. 1.

In order to maintain the main property of conditional probabilities, it is necessary that $\sum \hat{P}(w|w_{i-1}, ..., w_{i-n}) = 1, \forall w \in L$. Katz backoff accomplishes this by reserving part of the probability space for backing off. In our case, the reserve is similar to the Lidstone probability estimator (as seen on Eq. 8). The method adds a discount factor $\alpha|C_{S_n}^w|$ in the denominator of the MLE estimator, where α is the Lidstone parameter, S_n is a sequence of n words and $|C_{S_n}^w|$ is the number of different words that appear after the sequence of words. Once we do that, we have an discounted value D_{S_n} (Eq. 9) to be distributed in next

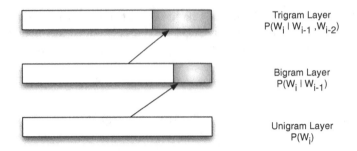

Fig. 1. Ilustration of the classic Katz backoff Model for n-gram language model

level of the backoff. This way, we define a value β (Eq. 10), that is a multiplier for the backoff probabilities into this reserved probability space. Notice that, for computing this value, we simply divide D_{S_n} by the amount not discounted in the following backoff level.

$$D_{S_n} = \frac{|C_{S_n}^w|}{f_{w_i-1,...w_n} + \alpha|C_{S_n}^w|} \tag{9}$$

$$\beta = \frac{D_{S_n}}{1 - D_{S_{n-1}}} \tag{10}$$

Finally, the general Katz backoff formula has two cases: (1) when the n-gram is in the dataset, we run a discounted probability estimation, (2) otherwise we call the backoff model and move it to the new probability space using the factor β (Eq. 11).

$$P(w_i|S_{n-1}) = \begin{cases} \frac{f_{w_i}}{f_{w_i,S_{n-1}} + \alpha|C_{S_n}^w|}, & \text{if } w_i,...w_{i-n} \in L \\ \beta P(w_i|S_{n-2}), & \text{otherwise} \end{cases} \tag{11}$$

In the last level (unigram level), we apply the classic Lidstone smoothing. Therefore, in a case where $w_i \notin L$, we would still get a non-null probability of $\hat{P}(w_i) = \alpha/(N + \alpha|C|)$. Notice that this only happens in the last level. In all other levels, Katz backoff prefers to search for a more general version of the n-gram in the following levels than atributing the same probability to all not found items.

4.2   Context-Enriched n-gram Language Model

n-gram models associated with Katz backoff are efficient methods for estimating the probability of sentences belonging to a language. However, based on Pragmatics theory, we believe that the decision whether a token w_i belongs to a language L does not depend only on the sequence of tokens that occurred before it in the sentence. We believe that context may be as important, or even more important, than this sequence in some cases. Therefore, we need new language models that do not rely solely on text but also on features used by human cognition to interpret sentences. In this section, we propose a novel language model that considers a contextual score as one of the priors in the conditional probability.

In this proposed language model, our goal is to estimate the probability $\hat{P}(w_i|C_j, w_{i-1}, ..., w_{i-n})$, where C_j is the contextual score of one of the proposed contextual models ($j = \{AKC, SC\}$, where AKC is the agent knowledge context and SC is the situational context). For maintaining the same probability estimation method (i.e. Katz backoff and Lidstone smoothing), we use C_j, a discretized value of the contextual score. This way, C_j can be considered as an artificial "word" that represents how likely it is for the message to be related to the subject or, in other words, to belong to the language. It is also important to stress that C_j is a message attribute. Therefore, its value is the same for all n-grams extracted in a single message.

The intuition behind the proposed attribute C_j is that we expect that some n-grams may belong to the language only in the presence of important contextual information. For instance, the trigram "And he scores" is much more likely to be related to the subject "football" when there is a football match going on. In this case, C_j would get a higher score during a football match and the language model would return a much higher value than for a contextualized n-gram. Another good example is when there is ambiguity in the n-gram. For example, if the trigram "NY is terrible" was uttered by a basketball fan, this would likely to be related to the NBA team, not to the city. In this scenario, C_j would be high when j is the AKC score and the language model would recognize that "NY is terrible" is a common trigram in this case. Unfortunately, adding the context into the language model is not trivial. In order to fully integrate it with classic n-gram model we had to modify Katz backoff.

To include context in the backoff process, we added new levels to the language model. The first level is the full contextual n-gram, for which the probability, $\hat{P}(w_i|C_j, w_{i-1}, ..., w_{i-n})$ is estimated. In the following level, we maintain the context C_j and remove the word token w_{i-n} from the prior, therefore we estimate the probability $\hat{P}(w_i|C_j, w_{i-1}, ..., w_{i-n-1})$. The contextual unigram is the last level in the contextual backoff. In this level, we maintain only the C_j in the prior, so the estimated probability is $\hat{P}(w_i|C_j)$. If this contextual unigram still cannot be found in the training set, we run the previous n-gram language model ignoring context.

Notice that, in this solution, we fall back to the traditional n-gram whenever context associated with text fails. However, it is possible to prioritize context when this happens. In order to accomplish that, we compute the probability distribution function (PDF) for the contextual scores and attribute it to the the token when we fail to find the contextual unigram. This way, instead of receiving $\hat{P}(w_i|w_{i-1}, ..., w_{i-n})$, which depends only on text, the backoff would get $PDF^j(C_j)$, which relies only on context.

The whole process is described in Fig. 2. In this figure, the shadowed part represents the discounted probability space that maps to the following level model. Remember that there are two options once the language model fails to find the contextual n-gram, one is attributing the contextual PDF (context emphasis) and the other backs off to the non-contextual n-gram (text emphasis). Both these options will be evaluated separately in Sect. 6.

Language models are efficient techniques for subject classification when there is no negative label. In this section, we defined methods for using extra-textual information in these models for performance improvement. In the next section, we show some characteristics of the dataset and demonstrate some of the defined hypotheses.

5 Dataset

For the experimental evaluation of the proposal in this paper, a dataset was gathered using Twitter's streaming API. This dataset comprises tweets which could be related to three popular American sports and their leagues: american football (NFL), baseball (MLB) and basketball (NBA). Our goal is to define if a given message references one of these subjects. For that, we define two subsets of messages: training stream and test stream. The training stream is used exclusively for model generation, while the test stream is the one we extract messages for labeling. In the next sessions, we describe the process of data gathering and some characterizations of the dataset.

5.1 Dataset Information

For evaluating our hypotheses and the proposed language model we use three training streams (one for each subject) and one test stream collected using Twitter's streaming API. The training streams were collected by keyword-selection approach. Since the keywords were manually choosen focusing on recall, they may include some ambiguous keywords. Consequently, the training stream may contain tweets that are not related to the subject.

For the test stream, we employed a user-selection approach. The set of users was chosen from each training stream according to the method we presented in the previous section. Then, we collected all tweets posted by these users.

All training streams were collected from October 17th, 2013 to November 4th, 2013. As for the test stream, we collected messages from October 27th, 2013 to November 4th, 2013. The user set used for the test stream was generated from tweets gathered from October 17th to October 27th. The number of collected tweets can be seen in Table 1.

During the test set period, the following events, related to the target subjects, took place:

– **Baseball (MLB):** World Series 2013 - St Louis Cardinals vs. Boston Red Sox - Matches happened on 10/23, 10/24, 10/26, 10/27, 10/28 and 10/30. In the end, the Red Sox won the series and became the 2013 champions.

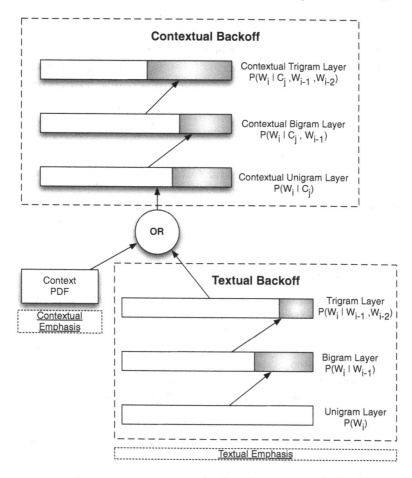

Fig. 2. Ilustration of the context-enriched language model backoff

- **American Football (NFL):** Mid Regular Season 2013 - Weeks 8 and 9 - All regular season matchups on NFL happen on Thursdays, Sundays and Mondays. Therefore, all matches happened on 10/27, 10/28, 10/31, 11/03 and 11/04.
- **Basketball (NBA):** Preseason and early season - Week 1 - There are matches every day since the start of the preseason.

It is expected that characteristics of these events have a major impact on contextual features and on the language model result. We will explore this in the next sections.

Labeling messages from the test dataset was not a trivial task. Since we wanted to improve the recall in messages that do not have textual references to the subject, we could not rely on traditional methods of annotation that only consider the text of the message. Therefore, to annotate a tweet we consider user profile, previous messages, time of post, mentions to other users and other con-

Table 1. Information about the collected streams

Dataset	Number of Tweets	Period
MLB training stream	20M	10/17/2013 - 11/04/2013
NFL training stream	16M	10/17/2013 - 11/04/2013
NBA training stream	19M	10/17/2013 - 11/04/2013
Test stream	100K	10/27/2013 - 11/04/2013

textual information. Of course, such annotations had to be performed manually, thus limiting the size of the training set.

In the next sections, we show some interesting information about the collected training stream and test stream. Since they were collected using different approaches, we discuss characteristics of these streams separately.

5.2 Training Stream

To generate the training stream R, we select all messages that contain at least one keyword $k \in K$ related to the subject S. The set of keywords K was chosen manually (for more information look at the appendice). To increase recall in the training stream, we chose to include many ambiguous keywords in K. Sport teams in the United States are traditionally named after a location and a nickname, for instance, "New York Giants" and "Baltimore Ravens". We chose to include in K, the location and the nickname of all teams separately (following the last example we would include "New York", "Giants", "Baltimore" and "Ravens") and some other keywords referencing to the sport (such as "touchdown", "nfl" and "field goal"). Notice that some nicknames and all locations are ambiguous. Fortunately, the proposed algorithms are robust towards ambiguity and are able to generate good contextual models even if many tweets in R do not reference subject S, as the experimental results in Sect. 6 show.

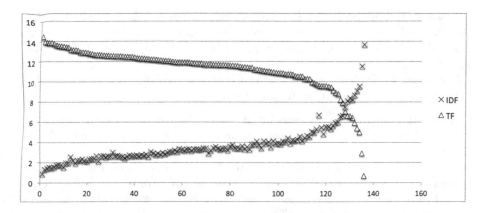

Fig. 3. Score of TF and IDF for each keyword ordered by the reverse TF order

Users Characterization. In Sect. 3, we proposed using weights inspired in TF-IDF for modeling the agent knowledge context (AKC). In our model, users are characterized by the set of keywords $k \in K$ used by them in R. The model proposes an analogy, in which users play the role of documents in traditional TF-IDF. In order to use TF-IDF-like weights for the AKC model, we need to show that the distribution of the TF component is similar to the IDF one [53]. It is intuitive to think that if one of them grows faster than the other, one term would be dominant. Figure 3 shows that this is not case in our database. In this figure, we plot the TF and IDF values for each keyword. We can see that many words have similar TF and IDF values in the center portion of the graph. This happens because of the similarity between keywords (most are team names and nicknames). The words with high IDF and low TF are the least ambiguous ones. We show an example of those in Table 2.

In Fig. 4, we plotted the TF-IDF weights for each keyword as a function of TF. This plot has a similar pattern to TF-IDF of terms in Web documents [53], in which the words that are given best TF-IDF scores are those with average TF. On Web documents, however, there is a concentration of terms with high frequency and low IDF that cannot be observed in our dataset because we never use stopwords in the K set. Therefore, we can see a low concentration of points in the right end of the horizontal axis. It is interesting to notice that this technique was able to give higher scores to keywords that are less ambiguous while giving lower score to more ambiguous words. For instance, terms such as "cheeseheads" and "diamondbacks" are almost exclusively used to refer indirectly to teams in NFL and MLB. Meanwhile, terms such as "NY" and "SF" are a lot more ambiguous. In Table 2 there are some examples of keywords that can be considered ambiguous and unambiguous and their IDF score. Moreover, both Fig. 3 and Table 2 are empirical demonstrations of Hypothesis 2, which posits that the ambiguity degree of a word is related with its frequency.

As expected, users behave differently in the training stream. One of the major differences is the number of keywords used during the analyzed period.

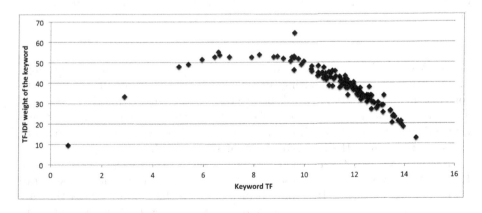

Fig. 4. Sum of all TF of a keyword by TF-IDF of that word

Table 2. Examples of ambiguous and unambiguous keywords

Unambiguous Keywords		Ambiguous Keywords	
Keyword	IDF	Keyword	IDF
cheesehead	52.51	SF	27.50
nyj	44.98	miami	21.19
white sox	52.75	boston	20.29
SF giants	52.70	NY	12.76
diamondbacks	53.76	new york	21.27

By grouping the users according to the variety of keywords used in the training stream we noticed that the size of these groups follows a power law. Therefore, few users post a wider variety of keywords while the majority used only a few.

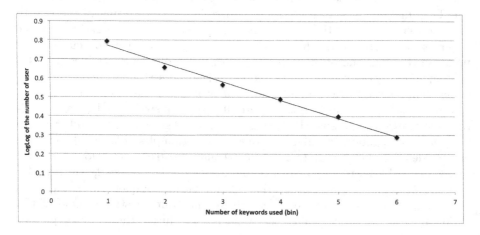

Fig. 5. LogLog of the number of users in each bin according to the keywords used by them

Time Frame Characterization. For estimating the likelihood of a message to be related to a subject during an arbitrarily defined frame of one hour, we consider the frequency of keywords posted during that timeframe. By Hypothesis 3, we believe that this likelihood is related to the amount of messages containing keywords posted during the timeframe. For penalizing ambiguous keywords, our proposed model uses the IDF score computed in the AKC model.

In Fig. 6, we plot the frequency for some keywords in each 1-hour interval (from Oct 28 to Nov 3) in the baseball dataset. It is easy to see that the keyword "NY" , which has a low IDF, only brings noise to the model. Words that are more related to the finals ("red sox" and "cardinal") clearly show two big peaks of frequency and a smaller one. These big peaks occurs exactly during the last two matches of the World Series. Therefore, this is a good example of Hypothesis 3: the increase in the frequency of these unambiguous keywords shows that there is an important event happening by that time. On the other hand we can see an ambiguous keyword ("boston") that followed the event peaks, but also had a really big peak on the last frames. Since "boston" has a low IDF, this probably offtopic peak does not have a major impact in the proposed model.

Figure 7 shows the Situational Context value (defined in Eq. 5) for each time frame in the MLB dataset. It is easy to see that it followed the peak pattern from Fig. 6, and also that noisy ambiguous words such as "NY" did not reduce the quality of the model. It is interesting to notice also that the higher values seen in this figure occur in late night periods with low traffic of messages in the United States. This is a good result, since we do not expect to have many messages related to the topic late at night. Another interesting observation is that our proposed situational context modeling is only sensitive for messages posted during major events. Therefore, when there is nothing relevant happening, we need to rely in other forms of context.

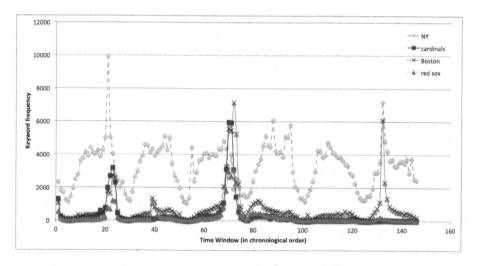

Fig. 6. Frequency of keywords in each one-hour time window of the training stream

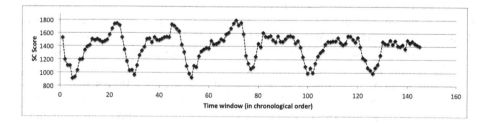

Fig. 7. Situational Context score in each time window of the training stream

5.3   Test Stream

By our main idea in this work, in a message we may only have an implicit reference to a subject. Therefore, simple keyword selection may get only a small subset of messages that reference the subjects. It is important for our purposes to have messages that make both kinds of references (implicit and explicit) to S in the test stream, since one of the major objectives of this work is to improve the recall in messages with implicit references to the subject. To collect the test stream we chose users that posted at least one message in the training stream.

Sampling the users to be collected is not an easy task. As shown in the previous section, the estimated interest degree of a user towards a subject follows a power law. Consequently, if we simply select users at random in the training stream, it is likely that we will mainly get users that are not really interested in the subject. This would result in a test stream with too few positive examples. For solving this problem, we discretized our user set in 10 buckets with different sizes, according to their AKC context. Then we sampled the same amount of users in each of them. IWe collected all tweets posted by a set of 300 sampled users.

Despite of the fact that we used different methods for collecting messages in each stream, we expect that all messages mentioning S implicitly or explicitly are under the same context and the same pragmatics effects. Consequently, we assume that using only messages that contain explicit references to S for training does not affect the performance of contextual models.

Characterization. One of the most important characteristics of the test stream is the presence of messages that are related to a given subject without having any of the keywords used for generating the training stream. In Table 3, we can see that more than half of the tweets that reference a subject, according to human annotators, did not include any of the keywords in K. This information shows us that relying on simple bag-of-words approaches for extracting subject-related messages may lead to low recall. Despite of that, we believe that most of messages that do not contain keywords are under the same context.

Figure 8 shows a scatter plot in which each point is a tweet, annotated as being related to a subject. The horizontal axis is the SC score of the message, while the vertical axis is the AKC score of the message. One important observation we can get is that there is no concentration of points with or without keywords in any region of the graph. This way we can conclude that, for our dataset, the contextual scores are independent from the presence of keywords. Another interesting side conclusion we can draw from this graph is that SC and AKC scores do not seem to be correlated.

In this section, we empirically demonstrated many hypotheses that we formulated throughout this paper. In the next section, we show the experimental results for the proposed language model and demonstrate the remaining hypotheses.

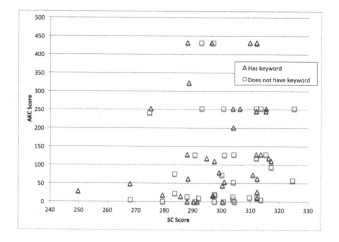

Fig. 8. Scatter plot showing the differences between AKC and SC

Table 3. Positive examples in the test stream

Subject	Positive Examples	% without keywords	% positives
Football	16	56.3	2.0
Baseball	14	71.4	2.0
Basketball	72	47.2	10.0

6   Results

In this section, we discuss the results of our proposed contextual-enriched language model for our three datasets. In these experiments, we add context scores for each tweet in both training and test stream according to our models. Once we have annotated all messages with contextual scores, we generate the language model using the training stream and then we evaluate it with messages in the test stream.

In this section, we denote messages referring to the target subject as positive examples, while the unrelated ones as negative. In the following sections, we discuss the evaluation method, and present experimental results.

6.1   Evaluation Workflow

For evaluating both the context-enriched language model and the pragmatic contextual models, we use the pipeline represented in Fig. 9. In this pipeline, we start with two separate datasets: training stream and test stream. As we argue in the Sect. 5, training stream is generated by keyword selection of messages while test stream is generated by all messages posted by a predefined set of users. The training stream is then used for generating the contextual model,

Agent Knowledge or Situational. Once the model is generated, we attribute the contextual score for each message in both training and test stream.

Then we sample examples fron the contextual-enriched training stream to generate the proposed language model. This sampling process is also being detailed in the next section. With the model generated we use the language model to give a score to each message in the contextual-enriched test stream. Once we have that score, we can run the following evaluation method to measure the effectiveness of our technique.

6.2   Evaluation Method

For evaluating the language model we used a test stream being all messages posted by a subset of users, as explained in the previous subsection. This way we are able evaluate messages unbiased by keyword-based recall approaches. However, the problem in this data stream is that positive examples are scarce. As showed previously (Table 3), less than 10 % of the messages are positive examples (i.e. related to the subject). In orded to overcome this issue, we use a ranking evaluation method [54] appropriate for this kind of data.

The evaluation method assumes that most of the unlabeled examples are negative. Given this assumption, we sample 100 unlabeled examples into a probe set. It is expected that most of those 100 examples are negative ones. Then, we add a positive example from the annotated ones into the probe set. Next, we rank all these examples according to the proposed model and baseline scores. Once this ranking is complete, we expect the positive example to figure at one of the top positions. We run this procedure 200 times for each positive example. Finally, we plot the average number of positives found up to the nth position. Since we want to improve recall, our goal is to have a higher percentage of examples seen for every n positions. In other words, we expect the line for the proposed method to be always over the baseline.

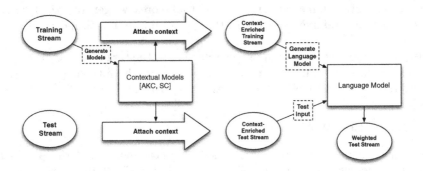

Fig. 9. Ilustration of the evaluation pipeline

6.3   Language Model Evaluation

Language models require a set of sentences (in our case, Twitter messages plus context) for training. It is expected that in the training stream we have many messages unrelated to the subject because of ambiguous keywords. To address this issue, we selected messages that are more likely to be positive in the training stream according to their contextual values and used them in model generation. In other words, we selected only the top two thirds of messages according to the Contextual Models and used them for language model generation. In the baseline, we randomly picked examples in the training stream so that we would get the same number of messages as in this sample. Therefore, in both training generation approaches we use the same number of messages to generate the language model.

For all experiments, we discretized the contextual scores given by the models in three bins that contain the same number of examples. We found out that dividing the dataset into more bins leads to worse results, since it fragments the training set.

We ran the experimental evaluation on five different methods. Each one uses context in a different way. The evaluated methods follow:

- **Baseline-n-gram (NG):** For this method we used the classic n-gram model generated with randomly picked examples from the training stream. Therefore, this method *completely ignores context* and *focuses only in the written message.*
- **Baseline-n-gram + contextual training (NGCT):** For this method we used the classic n-gram model generated with messages that were selected by their contextual values (the top two thirds of the messages, according to their contextual scores). This method *ignores context for model generation*. However, the training examples are supposedly better chosen than the previous method.
- **Context-enriched language model with textual emphasis (CELM-TE):** For this method we used the proposed context-enriched language model generated with messages that were selected by their contextual values (the top two thirds of the messages, according to their contextual scores). If there is not a contextual n-gram in a test example, this method backs off to the tradional n-gram and checks if it can be found there. Moreover, this method *only ignores context when contextual backoff fails.*
- **Context-enriched language model with contextual emphasis (CELM-CE):** For this method we used the proposed context-enriched language model generated with messages that were selected by their contextual values (the top two thirds of the messages, according to their contextual scores). If there is not a contextual n-gram in a test example, this method backs off to the raw contextual value. Therefore, this method *considers textual information only initialy* if associated with the context value.
- **Contextual Only:** For this method we ignore all message text. We *consider only the context value of the message.*

6.4   American Football Data

American football is the most influential of the American sports. With seasons scheduled from early September to late February, American football is known to have the most avid fan base. In the following experiments we test how important each context is for message understanding.

Agent Knowledge Context. Since the NFL football championship, during the analyzed period, was in mid-season, we believed that most of the tweets would be posted by fans and interested people. Moreover, we believe that in playoffs and finals (Superbowl), the championship may attract the attention of people that usually are not interested in this sport.

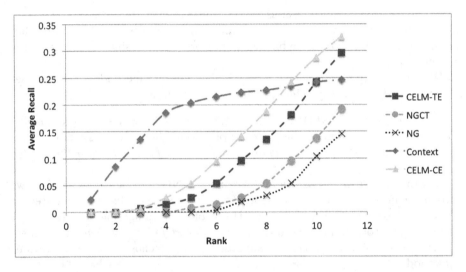

Fig. 10. American Football (AKC): average recall by ranking position considering all positive examples that do not contain keywords

As expected, we can see that context is really important, especially for messages without keywords (Fig. 10). In those cases, it is interesting to see that when we consider only context we get higher recall in the first positions. This means that there are some messages that have a high *AKC* score and bad textual information. Also, notice that the context-only approach stabilizes after 0.2 and from that point on the context-enriched language models perform better. We can conclude from this observation that around 20 % of the messages texts were poor and relying only on context leads to improved results. Also, we observe that in most of the messages a combination of context with written message provides better recall.

Analyzing results from the dataset with all positive messages (with and without keywords, Fig. 11) we can see a slight improvement of context-enriched language models over the baseline. Messages that already contain keywords are more likely to have strong text related to the subject. Therefore, contextual models add more information in this case. Since our goal is to increase the recall of messages, this result is less important than the last one, because, realistically, we would already have all messages that contain a given set of keywords. However, it is good to see that we do not have a worse result than the baseline even when text is supposedly strong.

Situational Context. All matches in the NFL championship are traditionally concentrated in a few days of the week (Thursday, Sunday and Monday). However, during the analyzed period we did not have any highly localized temporal event such as the Superbowl, playoff matches, or important news. Therefore we expected a result slightly worse than *AKC* Context.

The results, shown in Figs. 12 and 13, followed our expectations. The recall for messages without keywords follows a pattern similar to the *AKC*. However, we got better recall ratios at rank 10, which means that the Situational Context is slightly better than the *AKC* for the NFL during the analyzed period. This is an interesting conclusion that leads us to think that we can not rely on a single contextual source all the time. Since human cognition usually chooses between several context sources in order to decode the actual meaning of an utterance, we cannot assume that our simplified contextual model is able to make definitive decisions considering a single one.

For the messages with and without keywords (Fig. 13). We can see that contextual information improved recall about 5 % over the baseline. Once again, messages with keywords have better textual information, reducing the recall improvement in the proposed language model. Despite that, we can see that the

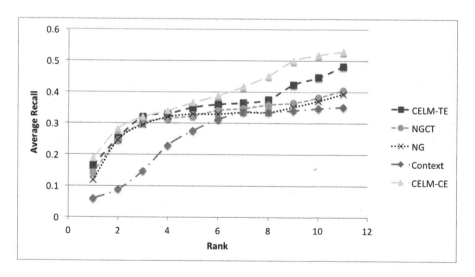

Fig. 11. American Football (AKC): average recall by ranking position considering all positive examples

contextual model with context backoff achieved better results. This is because messages with good textual results were already retrieved in the first positions.

Summary. Another important observation from this result is that the top messages retrieved considering only *SC* and *AKC* are different one from the other. This shows the complementariness of the proposed model. The top examples in *SC* were clearly related to matches, for instance "Golden Tate is THAT GUY!!", while *AKC* examples are more related to news and NFL-related histories, such as "Did LZ just correct AC again? "The old Robert Horry v Peyton Manning" comparison".

A conclusion that we can draw from this set of experiments is that there are some examples in the dataset that have really strong text. This commonly happens when we have a tweet that was retweeted many times. When this happens, we have good recall in the baseline for the first positions of the ranking (as seen on Fig. 13). However, the contextual model is more general and gets better results for the following positions.

6.5   Baseball Data

Baseball is a really important sport in the U.S., especially because of the time of the year in which the season happens. The MLB championship occurs mostly during summer, and for the majority of the season it does not compete with other American sports. Since a team plays more than 150 matches during the season, the importance of a single match event is low. However, the finals (World Series) happened during the analyzed period, and these were the most important

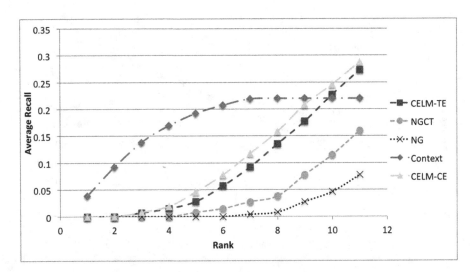

Fig. 12. American Football (SC): average recall by ranking position considering all positive examples that do not contain keywords

matches of the season. We explore the impacts of context during this specific period in the next results.

Agent Knowledge Context. During the finals, it is common for people who are not usually interested in baseball to post messages about the subject. Therefore, many positive examples were posted by users with average to low AKC. This generates low recall for the context-exclusive model. However, we can see that context-enriched language models achieved good results (Figs. 14 and 15) with recall ratios 8% to 10% better than the baseline at position 10.

The poor performance of the context-exclusive model and the good results of the context-enriched language model means that the activity of users with average AKC actually helped the proposed language model to get better results. Therefore, the language used by those users was actually a good representation for subject related messages, in that moment. It is interesting to notice that despite many positive examples that were posted by non-authorities, those users still had an important role for improving in our context-enriched language model.

Situational Context. Since the finals are the most important event of the year, we expected that the Situational Context would perform especially well. The results have shown in (Figs. 16 and 17) we would get a higher recall at rank 10 if we considered only the contextual information. It is interesting to notice that in cases in which we have a really strong context, text may reduce the model's performance. Still, we notice that in Fig. 17 the proposed model still improves the baseline by 5%.

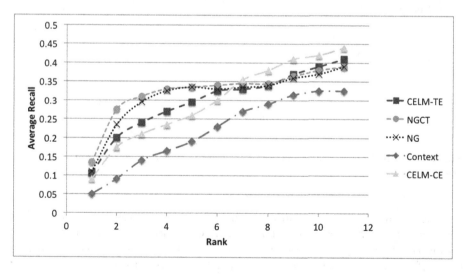

Fig. 13. American Football (SC): average recall by ranking position considering all positive examples

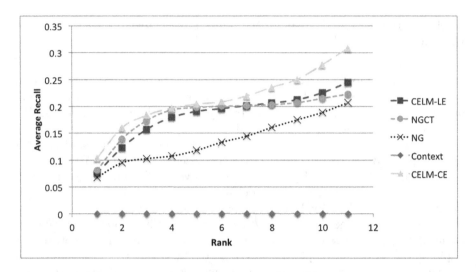

Fig. 14. Baseball (AKC): average recall by ranking position considering all positive examples that do not contain keywords

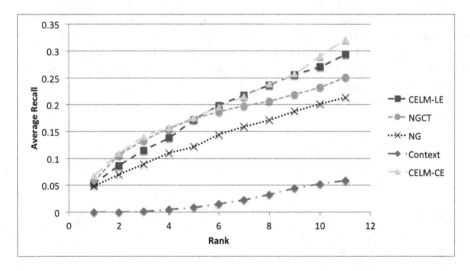

Fig. 15. Baseball (AKC): average recall by ranking position considering all positive examples

Another interesting observation is that the contextual-exclusive model gets low recall in the first positions. This can be explained by the number of posts during the finals that were not related at all with baseball. Despite the importance of the event, there are still many posts that are unrelated to the subject during the match. Those messages may frequently get the top positions.

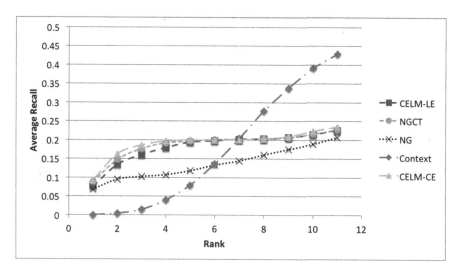

Fig. 16. Baseball (SC): average recall by ranking position considering all positive examples that do not contain keywords

Summary. The baseball dataset had the unique characteristic of being gathered during the finals. This enabled us to analyze the results in cases where there is a very strong situational context. A large number of people were interested in the final matches and commented about the outcome right after each one was over. This generated an overly strong situational context, and in this case the text actually reduced the performance of the proposed language model. In this cenario it would be more interesting to have a model that gave less importance to the text.

Another interesting observation from these experiments is that even when authorities are not the only ones posting messages related to the subject, we can still get good recall improvements from context-enriched language models. This happens because the vocabulary of all users related to the subject was similar. Therefore, learning from users with average AKC generated a good language model. A good example of the kind of language that was learned by the contextual model is in this tweet: "RT BrianCostaWSJ: they're pulling tonight's Silverado Strong promotion.". The MLB officials decided to veto the polemic advertisement "Silverado Strong", which played with Red Soxs slogan "Boston Strong". Since many users with average and high AKC were commenting on this decision, the language model worked well despite that this specific message was not posted by a user with high AKC.

6.6   Basketball Data

Basketball is one of the most popular American sports outside of the U.S. It also is known to have the highest engagement levels in Twitter among all American sports[7]. It is not only the fans activity that is famous in NBA, the league is also

---

[7] http://mashable.com/2013/04/25/nestivity-engaged-brands/.

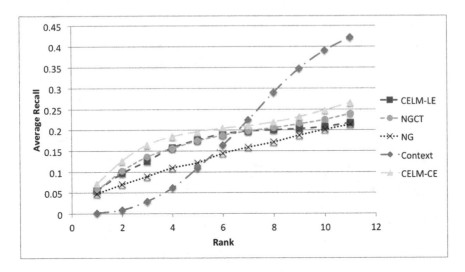

Fig. 17. Baseball (SC): average recall by ranking position considering all positive examples

known for having matches every day. All these factors combined allowed us to have more messages related to basketball than to all other sports, even when the analyzed period is mostly during preseason.

Agent Knowledge Context. The NBA, during the analyzed period, was under the preseason. During this period matches do not affect the final season score, they are mostly for practice. Consequently, we expected that most of the messages related to this subject in our test stream would be posted by users that are interested in the subject with high AKC. This is exactly what happened in both scenarios. In the one without keywords, we managed to achieve an improvement of over 13 % in recall. In the other one, we had a more modest improvement of only 4 %, because of the same reason, stated previously: messages with keywords have strong textual evidence that help baseline language models.

The exclusively contextual model, however, had bad results. Therefore, we can assume that we had a reasonable activity of non-authority users. This leads us to a conclusion similar to the one we had in the baseball dataset: even when context alone does not generate good recall, we get better results in our language models, which tend to be more general.

Figure 18, shows that, on the first ranking positions, all methods are equivalent. This happens because of the results with strong textual information. In those cases, the context, even if it is strong, adds few information for the top ranked messages.

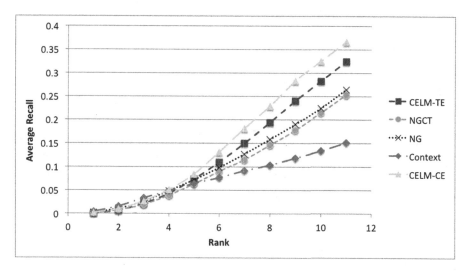

Fig. 18. Basketball (AKC): average recall by ranking position considering all positive examples that do not contain keywords

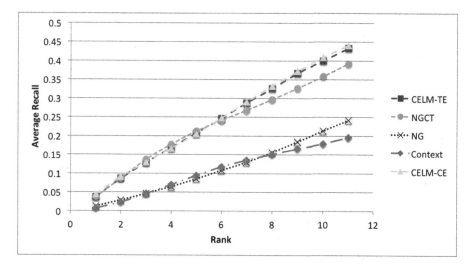

Fig. 19. Basketball (AKC): average recall by ranking position considering all positive examples

Situational Context. Since we have many unimportant NBA matches at each of the days analyzed, we expected a poor result of contextual models for this dataset. Results show that the situational context does not add a significant contribution to the n-gram baseline. When we consider all positively labeled messages, we can see that the addition of context in this case only increased the recall at rank 10 by 3 %. In the case where we consider only positive examples

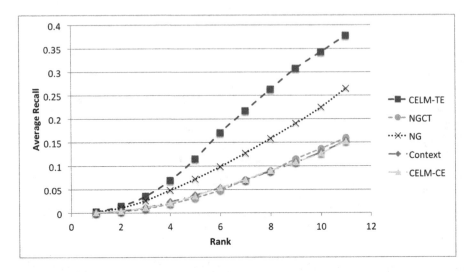

Fig. 20. Basketball (SC): Average recall by ranking position considering all positive examples that do not contain keywords

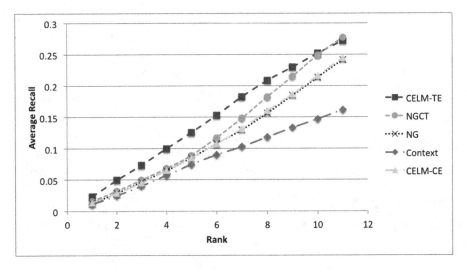

Fig. 21. Basketball (SC): Average recall by ranking position considering all positive examples

without keywords (Fig. 21), we can see that the n-gram baseline got the second best result. This means that context was not useful. On the contrary, context even reduced the performance of the model.

One interesting observation in both Figs. 20 and 21, is that these are the only cases in which the contextual enriched n-gram with textual backoff got a better result than the one with contextual backoff, showing that the extensive use of context, in this case, reduces the performance of the language. It is also important to emphasize that even with bad contextual information the proposed model was better than the baseline. This shows the robustness of the model and of the backoff technique.

In Fig. 21, the context-enriched n-gram with textual backoff gives a better recall ratio up to rank 9. After that, there is a draw with one of the baselines. It is an interesting conclusion that the context-enriched language models always get better results in the first positions, showing that they have the expected behavior of boosting the likelihood of messages that were uttered under higher contextual scores. This shows that our contextual models were actually able to identify messages that have a clear conversational implicature that indicates the reference to the message.

Summary. With a weak situational context and a stronger agent knowledge influence, the basketball dataset demonstrated that combining different contexts is an important future work, since the models perform differently in each dataset. However, despite of being worse, in both cases the context-enriched language model performed better than the baseline, demonstrating that even with a weak information gain, context is still a valuable information.

For the agent knowledge context, we achieved a big performance gain over the baseline, retrieving 13 % more examples. This is an improvement of about 50 % and can be explained by the characteristics of the dataset. Early in the season, the only people who post messages are passionate about this sport and are interested in the event. These results show us how context influences the way people communicate in social media.

7 Conclusions and Future Work

In this dissertation, we argued that considering only text is not enough for social media message comprehension. According to Pragmatics theory, speakers expect that their receivers consider contextual, or extra-textual, information for decoding and understanding messages. Given this assumption, speakers tend to omit from the uttered message information that they expect their receivers already hold. We showed that this implicit method of compression and channel optimization in human communication cannot be overlooked in Twitter messages, since about 70 % of subject-related messages in our test stream do not contain any keywords trivially related to the subject.

To address the insufficiency of text in the analyzed scenario, we proposed models based on TF-IDF for attributing scores to the relationship between context and target subject. Our models focus on two different extra-textual elements: the degree of interest of a user towards a subject (Agent Knowledge) and the likelihood that an important event related to that subject is happening (Situational). We demonstrated that these models are independent and that messages with and without keywords have similar contextual scores. These models had good performance, especially when there is a strong external signal, such as the MLB World Series matches. However, in many cases, they needed to be combined with text to achieve better results.

Our proposed technique was able to retrieve 100 % more messages in American Football dataset and more than 100 % more in messages in Basketball dataset for messages without any keywords. As for the Baseball dataset, the context is such a strong signal that ignoring text we were able achieve similar results. This result is expressive because we are increasing recall in messages that were once unretrievable by bag-of-words methods. It is also important to notice that this improvement demonstrates the impact of context in Twitter messages. Moreover, we also had improvement of recall in messages with keywords, showing that all messages suffers impact of context.

In order to simultaneously use text and pragmatic context, we proposed a novel language model that considers both the scores computed by our contextual techniques and the words in the message. We proposed two strategies to be used when a word cannot be found within the given contextual level: (1) ignore all extra-textual information and (2) use exclusively them. The results show that usually the second option achieves better results, except when we have very weak contextual information, such as in the basketball situational context. We also concluded that the performance of the proposed pragmatics models is not constant, as their results improve in the presence of real world events. Therefore, exploring this variation of importance in pragmatics discourse elements constitutes an interesting future work.

The contributions in this dissertation are not restricted to the pragmatics and language models. We also developed a novel framework for analyzing Tweets in a non-keyword driven retrieval approach. In this framework, we propose a new method for collecting messages without keywords in Twitter, given all API constraints. We also show the importance of not considering exclusively text in the labeling process. It is required for the annotator to check the profile, past messages and understand about the subject before labeling the message. Finally, we demonstrate the validity of our hypothesis that both messages selected by keyword and messages selected by user are under the same pragmatic context influence. This hypothesis opens the possibility to use two different streams for training and testing, which is an important perspective in this new retrieval approach.

8 Future Work

The verification of the high importance of non-textual elements in social media leads to a deterioration of the performance of classical keyword retrieval approaches. One important question that we need to answer in the future is how to evolve this classic retrieval model to one that is closer to the human intrisic cognition ability to interpret text. To reach this goal, we need to identify other relevant sources of contextual information in this form of communication. Then, another future work would be to create methods of combining all contextual information and text to improve recall. Finally, there is the need of adapting these techniques to work on a stream for creating a continuous subject-related messages source.

Indirectly, this work affects all techniques that rely solely on text for doing retreival and/or classifying social media messages. Since messages that contain keywords may be written differently from those that do not, we believe that techniques which use only text may get biased results. We believe that all techniques of sentiment analysis and trend detection in social media need to use contextual elements for retrieval and for their methods, otherwise they are falling in the pitfall of analyzing just an small subset of messages related with the desired target. Therefore, another possible future work is to create a tool for smart subject-related data gathering to be used by those techniques.

References

1. Community cleverness required. Nature, 455(7209), 1–1 (2008)
2. Calais Guerra, P.H., Veloso, A., Meira Jr, W., Almeida, V.: From bias to opinion: a transfer-learning approach to real-time sentiment analysis. In: Proceedings of the 17th ACM SIGKDD International Conference on Knowledge Discovery and Data Mining, pp. 150–158. ACM (2011)
3. Davis Jr., C.A., Pappa, G.L., de Oliveira, D.R.R., de L Arcanjo, F.: Inferring the location of Twitter messages based on user relationships. Trans. GIS 15(6), 735–751 (2011)
4. Gomide, J., Veloso, A., Meira Jr, W., Almeida, V., Benevenuto, F., Ferraz, F., Teixeira, M.: Dengue surveillance based on a computational model of spatio-temporal locality of Twitter. In: Proceedings of the 3rd International Web Science Conference, pp. 3. ACM (2011)
5. Levinson, S.C.: Pragmatics (Cambridge textbooks in linguistics). Cambridge Press, Cambridge (1983)
6. Yus, F.: Humor and the search for relevance. J. Pragmatics 35(9), 1295–1331 (2003)
7. Hanna, J.E., Tanenhaus, M.K.: Pragmatic effects on reference resolution in a collaborative task: evidence from eye movements. Cogn. Sci. 28(1), 105–115 (2004)
8. Cruse, D.A.: A Glossary of Semantics and Pragmatics. Edinburgh University Press, Edinburgh (2006)
9. Levinson, S.C.: Presumptive Meanings: The Theory of Generalized Conversational Implicature. MIT Press, Cambridge (2000)
10. Barbulet, G.: Social media- a pragmatic approach: contexts & implicatures. Procedia - Soc. Behav. Sci. 83, 422–426 (2013)
11. Pauls, A., Klein, D.: Faster and smaller n-gram language models. In: Proceedings of the 49th Annual Meeting of the Association for Computational Linguistics: Human Language Technologies - vol. 1, pp. 258–267, Stroudsburg, PA, USA, Association for Computational Linguistics (2011)
12. Saluja, A., Lane, I., Zhang, Y.: Context-aware language modeling for conversational speech translation. In: Proceedings of Machine Translation Summit XIII, Xiamen, China (2011)
13. Ifrim, G., Bakir, G. and Weikum, G.: Fast logistic regression for text categorization with variable-length n-grams. In: Proceedings of the 14th ACM SIGKDD International Conference on Knowledge Discovery and Data Mining, pp. 354–362. ACM, New York, NY, USA (2008)
14. Kurland, O., Lee, L., Hyperlinks, P.W.: Structural reranking using links induced by language models. ACM Trans. Inf. Syst. 28(4), 18:1–18:38 (2010)

15. Cavnar, W.B., Trenkle, J.M.: N-gram-based text categorization. In: Proceedings of 3rd Annual Symposium on Document Analysis and Information Retrieval, SDAIR-94, pp. 161–175 (1994)
16. Erkan, G.: Language model-based document clustering using random walks. In: Proceedings of the Main Conference on Human Language Technology Conference of the North American Chapter of the Association of Computational Linguistics, pp. 479–486, Stroudsburg, PA, USA, Association for Computational Linguistics (2006)
17. Peng, F., Schuurmans, D., Wang, S.: Augmenting naive bayes classifiers with statistical language models. Inf. Retrieval 7(3–4), 317–345 (2004)
18. Hayes, P.J., Knecht, L.E., Cellio, M.J.: A news story categorization system. In: Proceedings of the Second Conference on Applied Natural Language Processing, pp. 9–17, Stroudsburg, PA, USA, Association for Computational Linguistics (1988)
19. Yang, Y., Pedersen, J.O.: A comparative study on feature selection in text categorization. In: Proceedings of the Fourteenth International Conference on Machine Learning, pp. 412–420, San Francisco, CA, USA, Morgan Kaufmann Publishers Inc. (1997)
20. Mishne, G.: Blocking blog spam with language model disagreement. In: Proceedings of the First International Workshop on Adversarial Information Retrieval on the Web (AIRWeb) (2005)
21. Mishne, G.: Experiments with mood classification in blog posts. In: Proceedings of ACM SIGIR Workshop on Stylistic Analysis of Text for Information Access (2005)
22. Androutsopoulos, I., Koutsias, J., Chandrinos, K., Paliouras, G., Spyropoulos, C.: An evaluation of naive bayesian anti-spam filtering. In: Proceeding of the Workshop on Machine Learning in the New Information Age (2000)
23. Drucker, H., Wu, D., Vapnik, V.N.: Support vector machines for spam categorization. IEEE Trans. Neural Netw. 10(5), 1048–1054 (1999)
24. Joachims, T.: Text categorization with support vector machines: learning with many relevant features. In: Nédellec, Claire, Rouveirol, Céline (eds.) ECML 1998. LNCS, vol. 1398. Springer, Heidelberg (1998)
25. Sebastiani, F.: Machine learning in automated text categorization. ACM Comput. Surv. 34(1), 1–47 (2002)
26. Guyon, I., Elisseeff, A.: An introduction to variable and feature selection. J. Mach. Learn. Res. 3, 1157–1182 (2003)
27. Schwartz, R.M., Imai, T., Kubala, F., Nguyen, L., Makhoul, J.: A maximum likelihood model for topic classification of broadcast news. In: Kokkinakis, G., Fakotakis, N., Dermatas, E. (eds.) Eurospeech. ISCA (1997)
28. Natarajan, P., Prasad, R., Subramanian, K., Saleem, S., Choi, F., Schwartz, R.: Finding structure in noisy text: topic classification and unsupervised clustering. Int. J. Doc. Anal. Recognit. 10(3), 187–198 (2007)
29. Crammer, K., Dredze, M., Pereira, F.: Confidence-weighted linear classification for text categorization. J. Mach. Learn. Res. 13(1), 1891–1926 (2012)
30. Guan, H., Zhou, J., Guo, M.: A Class-feature-centroid classifier for text categorization. In: Proceedings of the 18th International Conference on World Wide Web, pp. 201–210. ACM, New York, NY, USA (2009)
31. Davis, A., Veloso, A., Da Silva, A.S., Meira Jr, W. and Laender, A.H.: Named entity disambiguation in streaming data. In: ACL 2012, pp. 815–824 (2012)
32. Li, Z., Xiong, Z., Zhang, Y., Liu, C., Li, K.: Fast text categorization using concise semantic analysis. Pattern Recogn. Lett. 32(3), 441–448 (2011)
33. Guo, Y., Shao, Z., Hua, N.: Automatic text categorization based on content analysis with cognitive situation models. Inf. Sci. 180(5), 613–630 (2010)

34. Qiming, L., Chen, E., Xiong, H.: A semantic term weighting scheme for text categorization. Expert Syst. Appl. **38**(10), 12708–12716 (2011)
35. Husby, S.D., Barbosa, D.: Topic classification of blog posts using distant supervision. In: Proceedings of the Workshop on Semantic Analysis in Social Media, pp. 28–36, Stroudsburg, PA, USA, Association for Computational Linguistics (2012)
36. Lao, N., Subramanya, A., Pereira, F., Cohen, W.W.: Reading the web with learned syntactic-semantic inference rules. In: Proceedings of the Joint Conference on Empirical Methods in Natural Language Processing and Computational Natural Language Learning, pp. 1017–1026, Stroudsburg, PA, USA, Association for Computational Linguistics (2012)
37. Li, C.H., Yang, J.C., Park, S.C.: Text categorization algorithms using semantic approaches, corpus-based thesaurus and WordNet. Expert Syst. Appl. **39**(1), 765–772 (2012)
38. Son, J.W., Kim, A. and Park, S.B.: A location-based news article recommendation with explicit localized semantic analysis. In: Proceedings of the 36th International ACM SIGIR Conference On Research and Development in Information Retrieval, pp. 293–302 (2013)
39. Machhour, H., Kassou, I.: Improving text categorization: A fully automated ontology based approach. In: 2013 Third International Conference on Communications and Information Technology (ICCIT), pp. 67–72 (2013)
40. Raghavan, S., Mooney, R.J., Hyeonseo, K.: Learning to read between the lines using bayesian logic programs. In: Proceedings of the 50th Annual Meeting of the Association for Computational Linguistics: Long Papers, vol.1, pp. 349–358. Association for Computational Linguistics (2012)
41. Lam, W., Meng, H.M.L., Wong, K.L., Yen, J.C.H.: Using contextual analysis for news event detection. Int. J. Intell. Syst. **16**(4), 525–546 (2001)
42. Yus, F.: Cyberpragmatics: Internet-Mediated Communication in Context. John Benjamins Publishing Company, Amsterdam (2011)
43. Susan C Herring. Computer-mediated discourse. The handbook of discourse analysis (2001)
44. Brody, S., Diakopoulos, N.: Cooooooooooooooollllllllllllll!!!!!!!!!!!!!: using word lengthening to detect sentiment in microblogs. In: Proceedings of the Conference on Empirical Methods in Natural Language Processing, pp. 562–570, Stroudsburg, PA, USA, Association for Computational Linguistics. (2011)
45. Howard, P.N., Parks, M.R.: Social media and political change: capacity, constraint, and consequence. J. Commun. **62**(2), 359–362 (2012)
46. Cha, Y., Bi, B., Hsieh, C.-C., Cho, J.: Incorporating popularity in topic models for social network analysis. In: Proceedings of the 36th International ACM SIGIR Conference on Research and Development in Information Retrieval, pp. 223–232 (2013)
47. Grice, P.: Syntax and semantics. 3: speech acts. In: Cole, P., Morgan, J.L. (eds.) Logic and Conversation. Academic Press, New York (1975)
48. Hirschberg, J.: A theory of scalar implicature. PhD thesis, University of Pennsylvania (1985)
49. Attardo, S.: Violation of conversational maxims and cooperation: the case of jokes. J. Pragmatics **19**(6), 537–558 (1993)
50. Eisterhold, J., Attardo, S., Boxer, D.: Reactions to irony in discourse: Evidence for the least disruption principle. J. Pragmatics **38**(8), 1239–1256 (2006)

51. Silva, I.S., Gomide, J., Veloso, A., Meira Jr, W. and Ferreira, R.: Effective senti-
    ment stream analysis with self-augmenting training and demand-driven projection.
    In: Proceedings of the 34th International ACM SIGIR Conference on Research and
    Development in Information Retrieval, pp. 475–484. ACM, New York, NY, USA
    (2011)
52. Phuvipadawat, S., Murata, T.: Breaking news detection and tracking in Twit-
    ter. In: Web Intelligence and Intelligent Agent Technology (WI-IAT), pp. 120–123
    (2010)
53. Baeza-Yates, R., Ribeiro-Neto, B.: Modern Information Retrieval. Addison-Wesley
    Longman Publishing Co. Inc, Boston (1999)
54. Cremonesi, P., Koren, Y., Turrin, R.: Performance of recommender algorithms on
    top-n recommendation tasks. In: Proceedings of the Fourth ACM Conference on
    Recommender Systems, pp. 39–46. ACM, New York, NY, USA (2010)

Improving Open Information Extraction for Semantic Web Tasks

Cheikh Kacfah Emani[1,2]($\boxtimes$), Catarina Ferreira Da Silva[2], Bruno Fiès[1], and Parisa Ghodous[2]

[1] Université Paris-Est, Centre Scientifique et Technique du Bâtiment (CSTB), Champs-sur-marne, France
{cheikh.kacfah, bruno.fies}@cstb.fr
[2] Université Lyon 1, LIRIS, CNRS, UMR5205, 69622 Villeurbanne, France
{cheikh.kacfah-emani, catarina.ferreira, parisa.ghodous}@univ-lyon1.fr

Abstract. Open Information Extraction (OIE) aims to automatically identify all the possible assertions within a sentence. Results of this task are usually a set of triples (subject, predicate, object). In this paper, we first present what OIE is and how it can be improved when we work in a given domain of knowledge. Using a corpus made up of sentences in building engineering construction, we obtain an improvement of more than 18 %. Next, we show how OIE can be used at a base of a high-level semantic web task. Here we have applied OIE on formalisation of natural language definitions. We test this formalisation task on a corpus of sentences defining concepts found in the pizza ontology. At this stage, 70.27 % of our 37 sentences-corpus are fully rewritten in OWL DL.

1 Introduction

In recent years, researchers have tackled the problem of Open Information Extraction in different manner: from machine learning [11] to the exploitation of sentence structure [3,10]. This last type of approaches obtains the best results. Unfortunately, their OIE-tools (exploiting grammatical dependencies [10] and syntactic tree [3]) sometimes output incorrect tuples. These wrong extractions are mainly due to parsing errors. Indeed, these approaches take advantage of the syntactic tree or grammatical dependencies provided by a parser. Consequently, a good way to improve Open Information Extraction is to handle parsing errors before the extraction stage itself. To achieve this goal, we have decided to *handle multi-word expressions* (MWE). A MWE is a phrase, made up of a set of words, which has a precise meaning and is unbreakable. "MWE-errors" represent more than 45 % of parsing errors. We propose an algorithm to shorten multi-word expressions (Sect. 3). We evaluate the algorithm on sentences targeting the domain of building engineering construction, and show how it outperforms existing approaches (Sect. 5.1).

Now that we provide a tool which has the ability to split a complex sentence into a set of simple triples we can use it to accomplish more high level task. We illustrate this in an automatic Natural Language (NL) to *Web Ontology*

N.T. Nguyen et al. (Eds.): TCCI XXI, LNCS 9630, pp. 139–158, 2016.
DOI: 10.1007/978-3-662-49521-6_6

Language Description Logics OWL DL [1] conversion process. Here NL sentences are definitions of concepts found in an existing ontology. The goal is to help ontology designers to acquire automatically the OWL DL expression of a concept whose definition is expressed in natural language. For instance when defining a *Spiccy Pizza* through the sentence "*A spiccy pizza is any pizza that has a spicy topping*" the formalisation task aims to provide the following OWL DL expression:

$$\text{SpiccyPizza} \sqsubseteq \text{Pizza } and \text{ (hasTopping } some \text{ SpiccyTopping)}$$

This result is built from entities (Pizza, SpiccyPizza, SpiccyTopping and hasTopping) of the pizza ontology[1]. On the contrary of approaches like [16, 23, 24], we do not create new entities from scratch. Our approach has the advantage to avoid the proliferation of entities and OWL DL expressions by taking advantage of the current state of the ontology. To obtain the formal expression of a sentence, we first extract all the assertions it contains using OIE. Next the challenge is to rewrite each triple as an OWL expression. Finally, all the expressions need to be recombined to obtain the final expression intended by the original definition. A preliminary evaluation of this approach has been made on a corpus of definitions of the pizza ontology. This corpus has 37 definitions from which 26 are fully and correctly formalised by our tool.

The key contributions of this work include:

- A betterment of Open Information Extraction (OIE) when taking into account domain terminology (Sect. 3)
- The proposal of a straightforward approach to provide a formal expression, in OWL DL, of a natural language definition (Sect. 4.1). This approach does not require any learning or external resource.
- The proposal of a formalisation approach which avoids an anarchic growing of new entities (Sect. 4.3). Indeed, the final expression, aligned with an existing domain ontology, uses *only* entities found within this ontology.
- A formalisation process which takes into consideration all the piece of information appearing in sentences (Sect. 4.2). This is done by taking advantage of OIE to ensure the grabbing of all the pieces of information.
- An original approach which merges all the triple extract from the sentence to obtain a single and coherent expression (Sect. 4.4).

The rest of the paper is organised as follows: first, a state of the art on both the identification the formal expression of concepts and Open Information Extraction (Sect. 2). Next, details about our approach to improve OIE when aware of domain terminology (Sect. 3). Then we expose our approach to obtain automatically OWL DL expressions from NL definitions, taking advantage of the enhanced OIE method we propose (Sect. 4). Next, we evaluate the two main problems we tackled (Sect. 5). Finally, we analyse the preliminary results we obtain (Sect. 6).

---

[1] http://www.cs.ox.ac.uk/isg/ontologies/UID/00793.owl.

2   Related Work

Some tasks in the Semantic Web community take as input a sentence and need to deal with *all the pieces of information* contained within it. It is the case for example of Question Answering (QA). To be able to answer a question, a QA system needs to be able to decode all the chunks of information held by the sentence. This inescapable need of identifying all what is said by the sentence can be achieved by OIE and it is what we apply to provide the formal expression of NL definitions. To the best of our knowledge, it is the first approach where definitions in natural language are automatically converted in their corresponding formal expressions containing entities completely disambiguated. Unlike approaches presented in [7,16], we do not learn the desired expression from the ontology or the knowledge base. In this work, we derive it from the concept's definition. Compared with some approaches working with NL sentences, we do not exploit partial information like in [25] where the authors focus on subsumption and thus exploit the *"is a"* fragment of the definition. Indeed, we are able to identify cardinality and value restrictions and we take in consideration all the evidences available within the sentence. In the attempt to identify more complex OWL restrictions, the interesting work of Tsatsaronis and colleagues [20] aims to provide the exact property, within a given ontology, which links two previously labelled concepts. Unlike our approach, a label denoting a class in the targeted ontology, first needs to be manually assigned to concepts and at the end of the process one does not have a complete formal expression reflecting the idea of the input sentence. In addition, our approach is part of a process which avoids the proliferation of new entities like in [23,24] where new ones are proposed to formalise a sentence. In our approach, we find the most suitable entities in the existing domain ontology able to transcript the idea conveyed by the NL definition. We are able to enrich lightweight ontologies with high logical specifications. It enables to obtain complex formal ontologies which are thus suitable for powerful reasoning (subsumption detection and consistency checking).

As mentioned in the above paragraph, OIE can help us to identify all the pieces of information within definitions, hence we need to use the most accurate OIE-tool. During the recent years, many systems were developed to perform OIE. It is the case of ReVerb [11], OLLIE [17], ClausIE [10] and CSD-IE [3,4]. ReVerb by means of efficient heuristics, focused on *incoherent* and *uninformative* triples. Unfortunately, relations extracted by ReVerb were necessarily verb-based. This is the main reason why OLLIE, developed by the same group of researchers, was provided. In addition to be able to identify non verb-driven facts, OLLIE aims to provide the *context/condition*, if existing, in which the extracted fact can *be considered true*. These two previous tools are machine learning-based. The most recent approaches do not need any additional resource. They only exploit result of a standard parser. ClausIE uses grammatical typed dependencies and CSD-IE the syntactic tree of the input sentence. These two tools dissect each piece of the result they get from the parsing tool. Consequently, if a dependency is wrong or a sub-tree is incorrectly labelled in the syntactic tree these OIE-tools may provide inaccurate extraction. This is why we propose to make some preprocessing operations before OIE itself. The details of these tasks are given in the next section.

3   Handle Multi-word Expressions to Improve Open Information Extraction

Recent approaches of OIE, for instance ClausIE [10] or CSD-IE [3,4], *only* use syntactic pieces of information to identify informative triples from a sentence. Consequently, errors in analysis of sentences by Natural Language Processing (NLP) tools lead to major incorrect extractions in Open Information Extraction. In a sample set of sentences selected from various regulatory texts in the field of building engineering (see Sect. 5.1 for more details about this corpus), the percentage of errors due to MWE is 46.15 % using CSD-IE. To handle problems caused by MWE is thus a relevant way to improve result of IE. Our solution to improve the quality of results of OIE-tools is to handle MWE by means of this three-step approach: (*i*) the *detection* of MWEs within the sentences, (*ii*) their *shortening* and then the information extraction process is done within each resulting triple containing a shortened-MWE, (*iii*) this shortened-MWE is expanded to get back to its original form. This process is illustrated by Fig. 1 and detailed in the following subsections.

Step 1 - Detection of Multi-word Expressions. For us, a MWE is every phrase which the meaning will be modified (even become meaningless) by the addition or the deletion of any of its word. Consequently, a *domain term*, an *idiomatic expression*, a *phrasal verb*, a *named entity*, a *formula*, a *quotation* etc. is a MWE. These types of MWE make us foresee that MWE are more easily and *reliably* identifiable in a given domain. One can have also domain-independent terms that are not related to the field of study but are frequently found in the corpus. It is the case of operators (example: less than, less than or equal to, as much as), idiomatic expressions (example: "Loose your head", "Jump in feet first"), units of measurements, etc. So, a set of MWE in a precise domain can be made up of the *terminology of the field* and *frequent terms*. This last category of terms can be obtained by means of existing statistical methods and the help of human experts. At this stage we identify the MWE present in the original sentence. We thus have a *list of possible MWE* in our corpus (see Sect. 5.1 for more details).

Step 2 - Compression of a Multi-word Expression. The reason why precision of OIE-tools is affected by MWE is that the latter is considered by the former to be *non atomic*. Hence, to limit potential hazardous fragmentation of expressions, we propose to extract information from a new version of sentences where each MWE will have been replaced by a *shortened version*. So, now the question is: how do we get this short version of MWE? When trying to answer this question, we must have in mind that the shortened sentences must always be semantically and syntactically correct to be appropriately handled by OIE-tools. We propose the following steps for shortening a MWE (using its syntactic parse tree):

1. if the MWE is a *clause* (list of labels for clauses is available in [6]) or a *verb phrase, there is no shortening*;
2. else, if the MWE is a noun phrase, the first token *labelled noun* is considered to be the shortened version of the MWE;

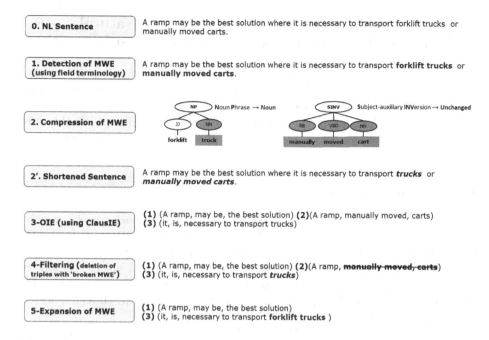

Fig. 1. An end-to-end example of Open Information Extraction when handling multi-word expressions.

3. else, we take the string provided by the smallest phrase[2] within the tree.

Let us note that some MWE will be short enough so that they will remain the same after the shortening. Although such MWE (like any other MWE) is considered to be atomic, this is important to have in mind because, if an OIE-tool breaks a MWE, the resulting triple will be incorrect.

After this stage, we now perform OIE itself, which is the **third step** (Fig. 1). This OIE is done by using existing OIE-systems. Consequently, the following steps come after OIE and take as input results of OIE, i.e. triples.

Step 4 - Filtering of Open Information Extraction Results. Earlier in this work, we have pointed out a set of things which degrades precision of OIE-tools. We have focused on the problematic role caused by multi-word expressions. Now, we use the only characteristic of MWE to finalise our OIE-process. Indeed, a MWE is *unbreakable*. Consequently, when a triple contains only a fragment of a MWE, it is considered as incorrect. This filtering is done before the expansion stage, so the MWE are in their "shortened" form.

Step 5 - Expansion of a Multi-word Expression. After the OIE has been done from the shortened version of the sentence, and te set of facts has been filtered, we now have to reconcile the remaining facts with the original (long) sentence. We then look into the list of the extracted facts to replace shorten version of MWE by their initial long form. This is the aim of this step.

---

[2] An exhaustive list of labels for phrases is available in the Penn Treebank [6].

4   Automatic Acquisition of Axioms from Natural Language Definitions via Open Information Extraction

In this section, we present our approach to formalise NL definitions (Sect. 4.1). The first step of this approach uses straightforwardly the improvements of our OIE method (Sect. 4.2). From OIE, we obtain a set of triples which are then formalised (Sect. 4.3). Then, we recombine all the formalised triple to obtain a single formal expression (Sect. 4.4). Finally, we provide a full example to illustrate our approach.

4.1   Overall Approach

Taking as input a sentence $\mathcal{S}$, a domain ontology $\mathcal{O}$, our approach aims to provide automatically the expression of the defined concept within $\mathcal{S}$ w.r.t to $\mathcal{O}$. We assume that the sentence follows the Aristotelian definition pattern [5]. In simple terms our definitions are made up of defined concept that is put in relation to one or more general concepts then, optionally, various precisions are added to these general concepts. Our formalisation approach is summarised as follows:

1. Structural Sentence Decomposition of $\mathcal{S}$ via OIE.
   The result of this step is a set of triples $\{\tau_i\}$ and their organisation (sentence structure) Σ.
2. For each triple τ_i
   (a) Identify and decode non domain terms.
   (b) Identify possible concept C_s and C_o in $\mathcal{O}$ using respectively the subject-part and the object-part of the triple.
       This identification is done by a string matching algorithm and we keep only concepts with the highest similarity score.
   (c) Identify the list $\{p_l\}$ of properties of $\mathcal{O}$ whose domain and range are *simultaneously compatible* with C_s and C_o.
   (d) Rank $\{p_l\}$ according to the matching score between each of its elements and the whole triple and select the top property.
   (e) Rewrite formally $(C \subseteq DL(\tau_i))$.
3. Link the triples as suggested in Σ.

The rest of this section details the steps of our approach.

4.2   Structural Sentence Decomposition

In our approach, we take advantage of Structural Sentence Decomposition using Open Information Extraction (OIE). This notion is exposed in details in [15]. Basically, when OIE is done, facts are *simply* extracted without any piece of information on how they are organised and linked. For instance, from the sentence "*If a building is intended to host the public then it should have two escapes or three main entrances*" we do not only have these three facts, triples $Fact_1$ <*A building, is intended, to host the public*>, $Fact_2$ <*it, should have, two escapes*>

and $Fact_3$ <*it, should have, three main entrances*>. Indeed, it is crucial to output how all these assertions are related as shown here: If($Fact_1$) then ($Fact_2$ or $Fact_3$). Since this organisation reflects the *structure* of the sentence, we call it *Sentence Structure* and we name it "*Decomposition*".

Here we use OIE with improvements bring by the handling of MWE as described earlier in this paper. A wrong information may lead to a wrong formalisation result. This is why we choose to deal with our improved OIE approach, despite some preprocessing steps before extraction itself. For the OIE step itself, we have chosen CSD-IE of Bast and Haussman [3]. First, let us mention that ClausIE [10] and CSD-IE are, to our knowledge, the best current OIE-tools. Secondly, characteristics of CSD-IE in comparison with ClausIE make the former more suitable for us than the latter. Indeed, CSD-IE was designed to provide triples with some quality aspects, mainly:

- *minimality*: triples should be small enough so that a more fine extraction cannot be made from them. In addition, in CSD-IE, *coverage* is a main concern. It means that there is an effort to make appear in the whole set of resulting triples each word of the original sentence at least once.
- *accuracy*: of course all systems, and thus ClausIE, are expected to be accurate. But, in CSD-IE, there is a given heuristic which has a worth for us: the predicate-part of their triples must only contain words which "belong to the verb". This is a guarantee of having a sort of *format* for resulting the triples. It will allow us to find string which may lead to a possible concept (Step 2.(b) of our approach in Sect. 4.1) *only* in the subject and object parts of the triple. We thus assume that verbs' (i.e. predicates of our minimal triple) contribution to the identification of a concept is negligible.

4.3 Processing of Each Triple

In accordance with the content of the previous section, our triples here are supposed to be minimal. Consequently, each of these triple will lead to a "*minimal restriction*" or a class subsumption. In OWL DL we have two main kinds of restrictions: *value* and *cardinality* constraints[3]. These restrictions in their *minimal* form *strictly* follow the templates presented below:

- Value constraints
  - owl:allValuesFrom: < *property* > only < *Class* >
  - owl:someValuesFrom: < *property* > some < *Class* >
  - owl:hasValue: < *property* > value < *Class* >
- Cardinality constraints
  - owl:maxCardinality: < *property* > max < *integer* >< *Class* >
  - owl:minCardinality: < *property* > min < *integer* >< *Class* >
  - owl:cardinality: < *property* > exactly < *integer* >< *Class* >

In these templates, words between ' <' and ' >' represent *slots* to be filled. The next paragraphs present elements related to the identification of the correct restriction and then its filling.

---

[3] http://www.w3.org/TR/owl-ref/#Restriction.

Handling of Non Domain Terms. When we look at our set of templates, we have some non domain terms i.e. terms not related to our domain ontology $\mathcal{O}$. These terms are OWL *key terms*. This operation consists of identifying the right non domain elements when taking as input a triple. As shown in existing works, mainly in *Question Answering* systems [21,22], they can be handled through a lexicon of non-domain terms. In this lexical resource, we have a set of NL expressions (e.g.: at least, higher than, uniquely, only etc.) and their formal equivalents, i.e. the exact OWL key term to which they refer and thus *the right template*. In this work, we have essentially taken advantage of existing lexicon of TBSL [21].

Concepts Identification. In our approach, we uppermost identify concepts instead of property. The reason is that, in one hand, we have some cases where there is not any hint to directly identify a property from the verb of the phrase. It mainly happens when the verb is a variation of the auxiliaries *be* or *have* (e.g.: <A wonderful pizza, has, a topping of tomato>). On the other hand, a given predicate can be expressed using different NL expressions (e.g.: A pizza *has/is made up of/contains* tomato). Here, we find a concept C_s using the subject-part and another one C_o using the object-part of the triple. C_o and C_s are found within $\mathcal{O}$.

Given an input string s and a domain ontology, we compare s with each *label* of all classes and individuals of the ontology[4]. For each comparison, we have a matching score and the class with the highest score is considered to be the concept (or individual) denoted by s. The challenge here is to have an *appropriate* metric for string alignment. Since in this task we are dealing with multi-token strings, we expect the following characteristics for the matching metric:

- the position of each token in corresponding string
- the similarity between each token of the two strings
- the editing distance (deletion, replacement, insertion) between the two strings.

In the literature, we have found a string similarity metric which fulfil all these features and which is called Liuppa [18]. It has been developed for ontology alignment at the terminological level. To handle the "token-level" of string comparison, designers of Liuppa first propose to rewrite each input string as a *"string of symbols"* as illustrate by Table 1.

We see that this replacement uses an existing string matching metric. After a set of experiments on our corpus, the Jaro-Winkler metrics [26] with a threshold equals to 0.85 gives the best results. The second and last step of matching performed by Liuppa is to provide the similarity between the two *"string of symbols"* (for the example exposed in Table 1, in this step one computes the similarity between $\alpha_1 \alpha_2 \alpha_3$ and $\alpha_3 \alpha_1$). Experimentally, we have chosen the Jaccard [12] metrics for this second matching. Indeed it is suitable for set-based similarity.

---

[4] With a large ontology, such comparison must take advantage of an index for the sake of scalability.

Table 1. From token to symbols as performed by `Liuppa` [18]. This operation takes as input two strings (here $S_1 = toppings\ of\ tomato$ and $S_2 = tomato\ topping$) and rewrite them as a set of symbols ($S_1 = \alpha_1\alpha_2\alpha_3$ and $S_2 = \alpha_3\alpha_1$). Each α_i represents two tokens similar over a given threshold and is seen as a character in the alphabet $\{\alpha_1,\ \alpha_2,\ \alpha_3\}$

		Token t	Most similar token t'	score(t,t')	Symbol
S_1	t_1	toppings			α_1
	t_2	of			α_2
	t_3	tomato			α_3
S_2	t_4	tomato	t_3 = tomato	1.0	α_3
	t_5	topping	t_1 = toppings	0.975	α_1

Because there is a lost of the order of the symbols when using `Jaccard`, we relax `Liuppa`(S_1,S_2) with `Jaro-Winkler` (S_1,S_2). The final similarity measure is given by:

$$Sim(S_1, S_2) = 0.75 \times \texttt{Liuppa}(S_1, S_2) + 0.25 \times \texttt{Jaro-Winkler}\ (S_1, S_2) \quad (1)$$

Properties Identification. Now we have possible concepts C_o and C_s, we have to provide the property which links them. Since we are linking out input triple with the domain ontology $\mathcal{O}$, this property should be found in the set of properties of $\mathcal{O}$. For this reason, we should first identify the set of properties $\{p_l\}$ in the schema of $\mathcal{O}$ which are *compatible* both with C_o and C_s. A property p_l is compatible with C_o and C_s if and only if:

$$Domain(p_l) \cap Hierarchy(C_s) \neq \emptyset\ and\ Range(p_l) \cap Hierarchy(C_o) \neq \emptyset \quad (2)$$

In Eq. 2, $Domain(p_l)$ and $Range(p_l)$ are respectively the range and the domain of p_l and $Hierarchy(x) = \{c, x\ \texttt{rdfs:subClassOf*}\ c\}$. The star (*) here denotes the *property path operator* for an arbitrary path of a length equals to 0 or more[5]. It allows us to select all the super-classes of x including x itself. $\{p_l\}$ can be obtained using the following SPARQL query:

```
PREFIX rdfs: <http://www.w3.org/2000/01/rdf-schema#>
PREFIX :<http://example.com/onto#>
SELECT DISTINCT ?p
WHERE  {:Cs rdfs:subClassOf* ?Dp . ?p rdfs:domain ?Dp.
         :Co rdfs:subClassOf* ?Rp . ?p rdfs:range ?Rp }
```

In this query :Co and :Cs are respectively URIs for the concepts C_o and C_s previously identified. When :Co or :Cs denotes an individual, the predicate `rdf:type` replaces `rdfs:subClassOf*` in the above query.

Selecting the Top Property. From the set $\{p_l\}$ of possible properties, we now need to rank them. This ranking will be done based on the strings provided by

---

[5] http://www.w3.org/TR/sparql11-query/#propertypath-arbitrary-length.

our input triple. Let us remind that using the subject-part and the object-part of the triple, we have identified the corresponding concepts C_s and C_o in the domain ontology $\mathcal{O}$. As first mentioned in Sect. 4.1 sketching our approach, we perform a matching between labels of each member of $\{p_l\}$ and the whole triple. The matching score is obtained using the following formula:

$$score(p_l, \tau_i) = \alpha.sim(p_l, pred(\tau_i)) + \beta.(score(p_l, subj(\tau_i)) + score(p_l, obj(\tau_i))) \quad (3)$$

In Eq. 3, α and β are two scalars used to weight the similarity between the current property and the predicate-part of the triple on one hand and the same property and the rest of the triple on the other hand. The relation between them is $\alpha + 2.\beta = 1$. We thus give more importance to the predicate than to the rest of the triple for property ranking. $subj(\tau_i)$ and $obj(\tau_i)$ represent the subject-part and the object-part of the triple τ_i. In this equation, sim represents the string similarity measure expressed in Eq. 1. In practice we have chosen $\alpha = 0.5$ and $\beta = 0.25$.

In brief, we assume that the predicate of a triple is pivotal for property identification. But in some cases, predicates are not informative enough. It is thus important to use the other part of the triple (subject and object).

Formal Rewriting of Triple. In the step 2.a of our approach, we have already identified the template of the restriction (and thus OWL key terms) and potential values. Now, using the concepts (C_o and C_s) and the property p_r, we just have to fill the selected template. We call $\mathcal{R}(\tau_i)$ (Restriction from τ_i) the output of this step.

4.4   Triples Join

Now that we have built our set of minimal restrictions, it is time to put them together to obtain the expression of our defined concept as stated by the input definition. Let us remind that our triples (still in NL) $(\tau_i)_{i=1}^{n}$ were obtained using a "Structural" Information Extraction approach, meaning that we know exactly how all these n triples are linked. As illustrated in Sect. 4.2, these links can be expressed by subordinate conjunctions (*if, then, while*, etc.), conjunctive adverbs (*unless, otherwise*, etc.) or by coordinating conjunctions (*and, or*). In this paper we will focus on cases where triples are linked by coordinating conjunctions (we discuss about other types of links in Sect. 5.2). We therefore notice that our triples are linked by *logical operators*. For us, negation is handled here at the *level of triple*. For instance, from the definition "*A vegetarian pizza is any pizza that does not have fish topping.*" we extract τ_1:-(A vegetarian pizza, is, any pizza) and τ_2:-(any pizza, does not have, fish topping). The structure of the sentence is $\Sigma = \tau_1$ **and** τ_2. We see that expressing τ_2 by (**not** (any pizza, has, fish topping)) does not have any impact on Σ. Consequently, the negation operator is not a possible link in Σ.

In this final step, we aim to answer the question raised by the following example.

We have the current input data:

- τ_1:-(A vegetarian pizza, is, any pizza)
  and thus $\mathcal{R}(\tau_1) = $:VegetarianPizza $\sqsubseteq$:Pizza
- τ_2:-(any pizza, does not have, fish topping)
  and $\mathcal{R}(\tau_2) = $ (not (:Pizza:hasTopping some:FishTopping))
- $\Sigma = \tau_1$ and τ_2
  and thus $\mathcal{R}(\Sigma) = \mathcal{R}(\tau_1)$ and $\mathcal{R}(\tau_2)$

How can we obtain:

$\mathcal{R}(\mathcal{S}) = $:VegetarianPizza $\sqsubseteq$ (:Pizza and (not(:hasTopping some :FishTopping)))

When we look at the expected result $\mathcal{R}(\mathcal{S})$, the task to perform seems to be a *join*, like in relational databases. Thus the name of *triples join*. In the above example, the *common attribute* for the join is the concept :Pizza. To obtain $\mathcal{R}(\mathcal{S})$, we will take advantage of join and boolean expressions simplification methods. In $\mathcal{R}(\Sigma)$ we already have some boolean operators (links between triples). Now, in the light of our approach to obtain them, for the sake of the current task, each $\mathcal{R}(\tau_i)$ is simply rewritten $C_s^i\, r_i\, C_o^i$. In this expression r_i stands for the restriction parameters[6].

For two triples τ_i and τ_j we perform simplification using these formulae:

$$\left(C_s^i\, r_i\, C_o^i\right) * \left(C_s^i\, r_j\, C_o^j\right) = C_s^i \sqsubseteq \left(r_i\, C_o^i * r_j\, C_o^j\right) \tag{4}$$

$$\left(C_s^i\, r_i\, C_o^i\right) * \left(C_o^i\, r_j\, C_o^j\right) = C_s^i \sqsubseteq \left(r_i\, (C_o^i * r_j\, C_o^j)\right) \tag{5}$$

In these formulae, the operator $*$ represents a logical operator (*and* or *or*). Moreover, as it is the case for join in relation database, a "*common*" operand to τ_i and τ_j is needed (in the following numbered list, we suppose $* = $ **and**[7]):

1. within Eq. 4, we have $C_s^i = C_s^j$ (the two restrictions concerned the same subject). The first member of this equation means "instances of C_s^i are individuals of the anonymous class $r_i\, C_o^i$ **and** instances of C_s^i are individuals of the anonymous class $r_j\, C_o^j$". Consequently, instances of C_s^i are member (subsumption) of both $r_i\, C_o^i$ and $r_j\, C_o^j$. In this case the join is similar to a *factorisation* of an arithmetic expression and we will refer to this first case by that name.

2. $C_s^j\, r_j\, C_o^j$ means "instances of C_s^j are individuals of the anonymous class $r_j\, C_o^j$". Now, suppose $C_o^i = C_s^j$ (Eq. 5). It implies that the restriction $r_j\, C_o^j$ has to be applied to C_o^i. This additional restriction for instances of C_o^i thus needs to be inserted in $\tau_i = (C_s^i\, r_i\, C_o^i)$. We call this equation where C_o^i is refined *refinement*.

---

[6] r_i is the subsumption or the set of elements of a more complex restriction (URI of the restriction property, OWL keywords for the type of the restriction, etc.) as explained in the introduction of Sect. 4.3.

[7] Only for better understanding. The choice of **or** would not have changed anything.

To evaluate the simplification of a complex expression, we use the Algorithm 1 below. In this algorithm we suppose the existence of the directed graph $G(V, E)$ thus defined:

- V is made up of all the set of distinct concepts found in $(\tau_i)_{i=1}^{n}$
- a directed edge (e_p, e_q) belongs to E if there exists a triple $\tau_i = (C_s^i \, r_i \, C_o^i)$ in $(\tau_i)_{i=1}^{n}$ with $C_s^i = e_p$ and $C_o^i = e_q$.

G is a dependency graph where a concept depends of another one if the former (in object position) helps to provide more information about the latter (in subject position). In G some vertices do not have any incoming edge and are element of what we call *Root*. Moreover, we can know at any time what is the triple which has lead to the edge $\left(C_s^i, C_o^i \right)$ in G (this information is kept during the built of G). We will use the function $\texttt{triple}(C_s, C_o, G)$ to denote this triple.

---

input : C, Σ, G

output: A triple $C \, r \, c_o$ which is the *factorisation* of all the triples having the concept C in subject position // Equation 4

$C_{objs} \leftarrow G.children(C);$
if $|C_{objs}| == 0$ **then** return $\varepsilon;$
// It means there is not any triple with C as subject, thus no
    factorization possible
else
> BoolLinks $\leftarrow$ BooleanLinks (Σ, C_{objs}) // ''Children'' of C (i.e. C_{objs})
>   and their links (boolean operators) in Σ (A directed edge in
> G corresponds to a triple in Σ)
> **for each** C_o in C_{objs} **do**
>> $\tau \leftarrow$ Triple $(C, C_o, G);$
>> $\tau' \leftarrow$ FactorizationRec $(C_o, \Sigma, G);$
>> BoolLinks $\leftarrow$ Refinement $(\tau, \tau',$ BoolLinks$);$// Equation 5
> **end**
> return Factorization $(C,$ BoolLinks$);$
end

Algorithm 1. Recursive factorization ($\texttt{FactorizationRec}$) algorithm

---

The *recursive* function $\texttt{FactorizationRec}$ depicted below "factorises" an input concept C. Before performing factorization itself ($\texttt{Factorization}$) the object-concepts in triples having C in subject position are firstly factorised and then refined ($\texttt{Refinement}$) in the aforementioned triples.

We call $\texttt{FactorizationRec}$ for each element of *Root*. In practise this set is usually a *singleton* containing the defined concepts. It is due to the Aristotelian form of the definition considered here.

4.5   An End-to-End Example

In this section we provide a full example of the formalisation of a NL definition.

Inputs: the defined concept *"American Pizza"* and its definition *"An american pizza is a pizza which has toppings of pepperoni, mozzarella and tomato."*

1. **Decomposition**
   - τ_1 (An american pizza, is, a pizza)
   - τ_2 (a pizza, has, toppings of pepperoni)
   - τ_3 (a pizza, has, toppings of mozzarella)
   - τ_4 (a pizza, has, toppings of tomato)
   - $\Sigma = (\tau_1)$ and τ_2 and τ_3 and τ_4
2. (a) **Identification and decoding of non domain terms**
   - τ_1 An american pizza $\underbrace{\text{is a}}_{\subseteq}$ pizza

   - No identification in the triples $\tau_2 - \tau_4$
   (b) **Concepts identification**
   - $(C_s, C_o)_{\tau_1} = $ (pizza:AmericanPizza, pizza:Pizza)
   - $(C_s, C_o)_{\tau_2} = $ (pizza:Pizza, pizza:PepperonniSausageTopping)
   - $(C_s, C_o)_{\tau_3} = $ (pizza:Pizza, pizza:MozzarellaTopping)
   - $(C_s, C_o)_{\tau_4} = $ (pizza:Pizza, pizza:TomatoTopping)
   (c) **Properties identification**
   - No identification in τ_1 since the predicate part of the triple has lead to a subsumption relation
   - In $\tau_2 - \tau_4$ we have the same set of properties { pizza:hasIngredient, pizza:isIngredientOf, pizza:hasTopping}
   (d) **Ranking**
   With the presence of words *has* and *toppings* the ordered list of properties (for triples $\tau_2 - \tau_4$) is (pizza:hasTopping, pizza:hasIngredient, pizza:isIngredientOf).
   Therefore the top property in this case is pizza:hasTopping.
   (e) **Formal rewriting** (when there is not any hint on the type of the restriction, we choose the *some values* restriction)
   - $\tau_1 \rightarrow$ AmericanPizza $\subseteq$ Pizza
   - $\tau_2 \rightarrow$ Pizza hasTopping some PepperonniSausageTopping
   - $\tau_3 \rightarrow$ Pizza hasTopping some MozzarellaTopping
   - $\tau_4 \rightarrow$ Pizza hasTopping some TomatoTopping
3. **Final linking w.r.t** Σ
   The dependency graph here is depicted by Fig. 2. The only root element (node without incoming edge) of this graph is the concept AmericanPizza. To factorize it, we will first need to do so with Pizza using $\tau_2 - \tau_4$. We obtain for the factorization of Pizza:

   Pizza $\subseteq$ ((hasTopping some PepperonniSausageTopping) and (hasTopping some MozzarellaTopping) and (hasTopping some TomatoTopping))

   Now, by taking the above triple $(C_s^j = $ Pizza, $r_j ='\subseteq'$, and $C_o^j = $ ((hasTopping some PepperonniSausageTopping) ...)) which adds more precision to the concept Pizza, we refine the same concept present in τ_1 and we obtain:

   AmericanPizza $\subseteq$ (Pizza and (hasTopping some PepperonniSausageTopping) and (hasTopping some MozzarellaTopping) and (hasTopping some TomatoTopping))

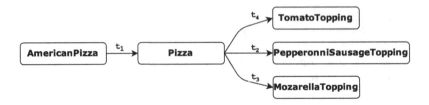

Fig. 2. Dependency graph from the triples provide in the example in Sect. 4.5

5   Overall Evaluation

The work we present in this paper tackle two main aspects. The first aspect is about the improvement of OIE using domain knowledge and the second one is an approach to provide the OWL DL expression of definitions in natural language. These two major tasks are evaluated into two following subsections. A proof of concept of the approach is available at http://tinyurl.com/ozkbmf3.

5.1   Multi-word Expressions Handling for Open Information Extraction

After making some statistics on the factors which lead to incorrect Information Extraction, we have decided to tackle the multi-word expressions-problem. The first step of the approach we propose is to identify them in the input sentence. Be able to perform such identification implies to have a list of possible MWE. That is why we hypothesize that the sentence describes the realities of a specific field of interest. Actually, to know that we are working in a specific domain implies to have a good idea of the terminology of this domain. For our evaluation, we have taken the list of terms (labels of the concepts in the field) as the set of our MWE. These terms have been obtained through a *key terms extraction process*. We have taken advantage of existing tools (Alchemy[8] in our case) to carry out this extraction. For this preliminary evaluation, our corpus is made up of 50 random sentences from documents about *fire safety* [8], *energy efficiency* [9] and *accessibility* [2]. Our list of MWE consists of result provided by Alchemy without terms containing a proper noun. Moreover, we have added units of measurement.

To perform OIE itself after preprocessing tasks, we have used ClausIE of Del Corro and Gemulla [10]. In addition, we compare our results to CSD-IE of Bast and Haussmann [3] and to the "original" version of ClausIE. Results are presented by Table 2.

CSD-IE performance in *this domain-specific* corpus (58.26 %) is less good than in "open datasets" like the Wikipedia (70.0 %) and New York Times dataset (71.5 %) [3]. The same remark can be made to ClausIE. But by handling the MWE-problem, we obtain 81.81 % of correct extractions (18.19 % of errors). We still have a certain number of errors. Some of these errors are caused by our handling of MWE as discussed in Sect. 6.1 and others errors come from OIE-tools we use at the extraction step itself.

---

[8] http://www.alchemyapi.com/.

Table 2. Results of the primary evaluation of OIE using `ClausIE` (by handling MWE -thus called `ClausIE-MWE`) and `CSD-IE` (without any preprocessing) on a corpus built from law in building engineering construction.

Tools	#extractions	#extractions-correct	#extractions-incorrect	#errors-due-to-MWE
`CSD-IE`	218	127 (58,26 %)	91 (41.74 %)	42 (46.15 %)
`ClausIE`	315	201 (63.80 %)	114 (36.20 %)	60 (52.63 %)
`ClausIE-MWE`	165	135 (81.81 %)	30 (18.89 %)	9 (30 %)

5.2 Automatic Formalisation Process of Definitions

Test Material. To the best of our knowledge, there is not any standard benchmark for automatic formalisation of NL definitions. Consequently we have made preliminary test on a well-known ontology from the SW community: the pizza ontology. The pizza ontology is very suitable for our task since it matches well our hypothesis: we have and existing schema with a given hierarchy, we also have a set of properties with their domain and range well defined, we have a set of new concepts and their definition to add to this pre-existant schema. We have taken the version 1.5 of the pizza ontology and gather their definitions (as stated in the official documentation [13]) when not provided in the `rdfs:comment` annotation of concepts. Since we have estimated that the 13 definitions we had after this operation were not enough for a real evaluation, we have gather more definitions on the Web. Indeed, there are many `:NamedPizza` is this ontology with no NL definition or comment. We look for possible definitions in Wikipedia and various restaurants sites[9]. In addition, these definitions are challenging because they do not follow the "classical" pattern denoted by **[Concept] <is a/are a> [Definition]** like in [23] or [19]. Indeed, the defined term is not always the introducer of the sentence (e.g.: *"Any pizza that has a spicy topping **is a spicy pizza**"*) or can be done using another verb (e.g.: *"A high calorie pizza **will be defined to be** any pizza ... "*). Moreover, since the pizza ontology example has been designed to illustrate the whole set of possible OWL restrictions, we have of course all these possibles restrictions expressed in NL definitions. At this stage we have a set of 37 definitions.

Results. The first step of our approach brings us to perform an Open Information Extraction (OIE). Of course, this operation does not always provide accurate results. An evaluation of OIE when taking into account domain knowledge (and thus multi-word expressions - MWE) is described in [14]. There, the accuracy is 81.81 % when using `ClausIE` for the extraction task itself. In this case, the domain knowledge (i.e. the terminology) taken as input was the set of labels of our entities. By taking into account this list of domain terms, we improve OIE from 75.2 % (when using directly `CSD-IE`) to 85.12 % as shown by Table 3.

---

[9] For example http://www.pizzaexpress.com/our-food/our-restaurant-menu/mains/, http://www.nutritionrank.com/, etc.

Table 3. Results of the evaluation of OIE when using `CSD-IE` (by handling MWE -thus called `CSD-IE-MWE`) and `CSD-IE` (without any preprocessing) on our set of definitions.

Tools	#Triples	#Triples-correct	#Triples-incorrect
`CSD-IE`	125	94 (75.2 %)	31 (24.8 %)
`CSD-IE-MWE`	121	103 (85.12 %)	18 (14.88 %)

Table 4. Results of the formalisation process

#Definitions	#Correct-Definitions-OWL	#Triples-correct	#Triples.-correct-OWL
37	26 (70.27 %)	103	82 (79.61 %)

For the extraction of restrictions itself, validation was done manually. Over the 37 expressions provided by the formalisation of our sentences, 26 (70.27 %) were correct (i.e. both triples in OWL DL and the final expression of the defined concept). From the 121 triples extracted (still in NL), 103 (85.12 %) were accurate. When formalising these triples, we get a precision of 79.61 % (82 triples). We remark that our matching process handle correctly *noise*[10] in NL expressions. These numbers are summarised in Table 4.

6   Overall Discussion

Preliminary evaluations we have performed both on improving OIE and automatic formalisation of definitions highlight some interesting points. They are discussed in the next two subsections.

6.1   About Open Information Extraction Improvement

We have seen that handle MWE, in a given domain, helps to improve OIE on sentences of that domain. However we see that our method to handle MWE has to be improved. Indeed 30 % of remaining errors after the shortening of MWE are due to this shortening operation. Indeed:

– When we choose the first noun of a *noun phrase*-MWE to replace this MWE it is not always its suitable representative. In practice, some nouns can sometimes be *tagged as verb* and thus a *potential predicate* (e.g.: "fire" in the expression *fire extinguisher*, "means" in the term *means of access*, etc.) and it can cause wrong extractions. Consequently, when we have more than a noun in a noun phrase, we must have more criteria to choose the representative.
– In some sentences, parsers correctly identify the prepositional modifiers of all verbs, nouns, adverbs, etc. Consequently the *presence of MWE is a priori not a problem* for OIE-systems. Unfortunately, the deletion of prepositions

---

[10] Concepts' tokens are usually surrounded by adjectives, adverbs, prepositions, etc.

(found for example in a *noun phrase*-MWE) during the shortening may lead to parsing errors. Indeed, parsers will try to identify new relations which may be wrong leading to incorrect extractions as illustrated below:

1. `Original sentence` : "A stair is a fixed means of access."
2. `Shortened version` : "A stair is a fixed means."
3. `OIE` : `CSD-IE`→(A stair, means, is) & `ClausIE`→(A stair, a fixed means)

One of the possible solutions to avoid the shortening of MWE to escape from their multi-word problem is to replace a MWE by a *synonym*. Ideally, this synonym should have fewer words (eventually a single word) than the original MWE. Such synonyms could be found in Linked Open Data or in lexical databases like Wordnet, etc.

6.2   Analysis of Our Automatic Formalisation Approach

In this section, we analyse the key steps of our approach in the light of the evaluation we have performed. In addition we highlight possible perspectives to this work.

Improving OIE. By using the compression of multi-word expressions we make a step in the direction of a more precise OIE-tool. But another weakness of OIE tools is enumeration detection. In a definition it is very common to enumerate items. The challenge for OIE-tools is to be able to decode correctly these items and to put them in the right place in a minimal <subject, predicate, object> representation. This task goes directly in hand with the identification of the structure of the sentence.

Triple Formalisation. We have made some (implicit) hypothesis for this step and they are discussed here.

1. In our approach we assume that each triple correspond to a restriction. In practice, it is not always the case. Some triples are not informative. For instance, for the sentence *"Sicilian pizza is a pizza prepared in a manner that originated in Italy."* we have the following output <A pizza, is prepared, in a manner> which does not have a correct formal expression w.r.t the pizza ontology. Such cases must be handled.

2. We do not take into account the fact that the concepts C_s and C_o could play opposite roles w.r.t the ontology. For instance, let us take the triple <Tomato, can be found, in Diavollo Pizza>. The restriction expected here is `DiavolloPizza` $\subseteq$ (`hasTopping` *some* `TomatoTopping`). Therefore, in the query to identify compatibles properties (Eq. 2) we may take this issue into account. Moreover, we may have a domain ontology where there are only properties from C_o to C_s.

3. We have performed a concept-driven approach (identification of concepts first and then their compatible properties). In some cases where the identification of the property (by means of a string matching) would have been correct, the concepts are wrongly identified thus leading to an incorrect property identification. We thus have to find hints, for each specific triple whether to have a property or concept-driven approach.

4. Another assumption made in this approach is the fact that C_s and C_o are directly linked. In other words, there is a *one length path* between them. Since we are in a perspective where the formalisation must be done taking into account the complexity of the schema of the ontology, we are not in principle aware of the existence of such a direct link. Thus, we must be able to handle paths made of many edges between concepts.

5. We have focused on object properties-based restrictions. But in practice we may find some restrictions based on data properties. We have in mind mainly the `owl:dataSomeValuesFrom` restriction which can be found for instance in the definition of a *High Calorie Pizza*:
   $\sqsubseteq$ `Pizza` *and* (`hasCalorificContentValue` *some integer*$[>= 400]$)

Going beyond Definitions. As early mentioned in Sect. 4.4, definitions can be done through sentences with a more complex structure than boolean connections between facts (e.g.: "*If* a pizza is made in Italy **then** it is an Italian pizza"). If we are able to join correctly the triples provided by complex NL definitions, it will open the way for a more general task on NL sentences: we mainly think about automatic formalisation of rules.

7 Conclusion

Open Information Extraction is an NLP task which aims to identify all the atomic facts within sentences. In this work we propose an approach to improve current OIE-systems overall performance with domain knowledge. This betterment is done without modifying the *code* of existing tools by handling multi-word expressions (domain terms) at the entry point (detection and shortening before extraction-itself) and the exit point (re-expansion of shortened expressions). This manner to deal with MWE allows us to improve OIE from more than 18 %. The ability of OIE to split a sentence into atomic and informative phrases is proposed to formalise automatically NL definitions in `OWL DL`. This formalisation process is straightforward and does not require any learning. In addition to subsumption detection, we are able to identify more complex `OWL DL` restrictions: cardinality and value restrictions. This formalisation task has been evaluated on a corpus made of sentences defining concepts of the pizza ontology. From the 37 sentences considered in the evaluation, 26 have been fully and accurately converted in an `OWL DL` expression. In the future, we must extend the set of possible restrictions actually considered in our approach. Moreover, we must enlarge the size of evaluation corpora. It may lead to some adjustments of some heuristics or weights for string similarity measures we used.

References

1. OWL Web Ontology Language Guide, February 2004. http://www.w3.org/TR/owl-guide/
2. American with Disabilities Act (ADA): 2010 ADA Standards for Accessible Design, September 2010. http://www.fire.tas.gov.au/userfiles/stuartp/file/Publications/FireSafetyInBuildings.pdf

3. Bast, H., Haussmann, E.: Open information extraction via contextual sentence decomposition. In: 2013 IEEE Seventh International Conference on Semantic Computing (ICSC), pp. 154–159. IEEE Computer Society (2013)
4. Bast, H., Haussmann, E.: More informative open information extraction via simple inference. In: de Rijke, M., Kenter, T., de Vries, A.P., Zhai, C.X., de Jong, F., Radinsky, K., Hofmann, K. (eds.) ECIR 2014. LNCS, vol. 8416, pp. 585–590. Springer, Heidelberg (2014)
5. Berg, J.: Aristotle's theory of definition. In: ATTI del Convegno Internazionale di Storia della Logica, pp. 19–30 (1982)
6. Bies, A., Ferguson, M., Katz, K., MacIntyre, R., Tredinnick, V., Kim, G., Marcinkiewicz, M.A., Schasberger, B.: Bracketing guidelines for treebank II Style Penn Treebank project. University of Pennsylvania 97 (1995)
7. Bühmann, L., Fleischhacker, D., Lehmann, J., Melo, A., Völker, J.: Inductive lexical learning of class expressions. In: Janowicz, K., Schlobach, S., Lambrix, P., Hyvönen, E. (eds.) EKAW 2014. LNCS, vol. 8876, pp. 42–53. Springer, Heidelberg (2014)
8. Building Safety Unit Tasmania Fire Service: Fire Safety in Buildings, obligaitions of owners and occupiers, August 2002. http://www.fire.tas.gov.au/userfiles/stuartp/file/Publications/FireSafetyInBuildings.pdf
9. California Energy Commission: 2008 Building Energy Efficiency Standards, for residential and nonresidential buildings (2008). http://www.energy.ca.gov/2008publications/CEC-400-2008-001/CEC-400-2008-001-CMF.PDF
10. Del Corro, L., Gemulla, R.: Clausie: clause-based open information extraction. In: Proceedings of the 22nd International Conference on World Wide Web, WWW 2013, International World Wide Web Conferences Steering Committee, Republic and Canton of Geneva, Switzerland, pp. 355–366 (2013)
11. Fader, A., Soderland, S., Etzioni, O.: Identifying relations for open information extraction. In: Proceedings of the Conference on Empirical Methods in Natural Language Processing, EMNLP 2011, Association for Computational Linguistics, Stroudsburg, PA, USA, pp. 1535–1545 (2011)
12. Hadjieleftheriou, M., Srivastava, D.: Weighted set-based string similarity. IEEE Data Eng. Bull. $33(1)$, 25–36 (2010)
13. Horridge, M., Jupp, S., Moulton, G., Rector, A., Stevens, R., Wroe, C.: A Practical Guide To Building OWL Ontologies Using Protégé 4 and CO-ODE Tools Edition1.2. The University of Manchester, Manchester (2009)
14. Kacfah Emani, C.H., Ferreira Da Silva, C., B., Ghodous, P.: Improving open information extraction using domain knowledge. In: Surfacing the Deep and the Social Web (SDSW), Co-Located with The 13th ISWC, October 2014
15. Kacfah Emani, C.H., Ferreira Da Silva, C., Fis, B., Ghodous, P., Khosrowshahi, F.: Structural sentence decomposition via open information extraction. In: 18th International Conference Information Visualisation (IV2014), July 2014
16. Lehmann, J., Auer, S., Bühmann, L., Tramp, S.: Class expression learning for ontology engineering. Web Semant. Sci. Serv. Agents World Wide Web $9(1)$, 71–81 (2011)
17. Mausam, S.,M., Bart, R., Soderland, S., Etzioni, O.: Open language learning for information extraction. In: EMNLP-CoNLL, pp. 523–534. Association for Computational Linguistics (2012)
18. Nguyen, V.T., Sallaberry, C., Gaio, M.: Mesure de la similarité entre termes et labels de concepts ontologiques. In: Conférence en Recherche D'information et Applications, pp. 415–430 (2013)

19. Sayah, K.: Automated Norm Extraction from Legal Texts. Master's thesis, Utrecht University, August 2004
20. Tsatsaronis, G., Petrova, A., Kissa, M., Ma, Y., Distel, F., Baader, F., Schroeder, M.: Learning formal definitions for biomedical concepts. In: OWLED (2013)
21. Unger, C., Bühmann, L., Lehmann, J., Ngonga Ngomo, A.C., Gerber, D., Cimiano, P.: Template-based question answering over RDF data. In: Proceedings of the 21st International Conference on World Wide Web, WWW 2012, pp. 639–648. ACM, New York (2012)
22. Unger, C., Cimiano, P.: Pythia: compositional meaning construction for ontology-based question answering on the semantic web. In: Muñoz, R., Montoyo, A., Métais, E. (eds.) NLDB 2011. LNCS, vol. 6716, pp. 153–160. Springer, Heidelberg (2011)
23. Völker, J., Hitzler, P., Cimiano, P.: Acquisition of OWL DL axioms from lexical resources. In: Franconi, E., Kifer, M., May, W. (eds.) ESWC 2007. LNCS, vol. 4519, pp. 670–685. Springer, Heidelberg (2007)
24. Völker, J., Rudolph, S.: Lexico-logical acquisition of OWL DL axioms. In: Medina, R., Obiedkov, S. (eds.) ICFCA 2008. LNCS (LNAI), vol. 4933, pp. 62–77. Springer, Heidelberg (2008)
25. Wächter, T., Schroeder, M.: Semi-automated ontology generation within obo-edit. Bioinformatics 26(12), i88–i96 (2010)
26. Winkler, W.E.: The state of record linkage and current research problems. Technical report, Statistical Research Division, U.S. Census Bureau (1999)

Searching Web 2.0 Data Through Entity-Based Aggregation

Ekaterini Ioannou[1]([✉]) and Yannis Velegrakis[2]

[1] Technical University of Crete, Chania, Greece
ioannou@softnet.tuc.gr
[2] University of Trento, Trento, Italy
velgias@disi.unitn.eu

Abstract. Entity-based searching has been introduced as a way of allowing users and applications to retrieve information about a specific real world object such as a person, an event, or a location. Recent advances in crawling, information extraction, and data exchange technologies have brought a new era in data management, typically referred to through the term Web 2.0. Entity searching over Web 2.0 data facilitates the retrieval of relevant information from the plethora of data available in semantic and social web applications.

Effective entity searching over a variety of sources requires the integration of the different pieces of information that refer to the same real world entity. Entity-based aggregation of Web 2.0 data is an effective mechanism towards this direction. Adopting the suggestions of the Linked Data movement, aggregators are able to efficiently match and merge the data that refer to the same real world object.

Keywords: Semantic web · Data integration · Semantic data management

1 Introduction

1.1 Challenges

Implementing entity-based aggregation to support entity search, poses a number of challenges due to the peculiarities of the modern web data, and specifically of that of Web 2.0 [3]. In particular, integration support needs to provide some *coherence* guarantees, i.e., to ensure that it can detect whether difference pieces of information in different sources representing the same real world object, are actually linked. It is not rare the case in which different sources contain quite different information about the same entity.

To successfully provide the above functionality, the aggregator needs first to cope with *heterogeneity*. Web 2.0 applications have typically a large amount of user-generated data, e.g., text, messages, tags, that are highly heterogeneous either by nature, or by design. For instance, DBPedia is based on RDF data and utilizes its own ontology, whereas Wikipedia adopts a more loose schema binding.

© Springer-Verlag Berlin Heidelberg 2016
N.T. Nguyen et al. (Eds.): TCCI XXI, LNCS 9630, pp. 159–174, 2016.
DOI: 10.1007/978-3-662-49521-6_7

The aggregator should be able to deal with a wide variety of entity descriptions ranging from keyword style entity requests to more structured and detailed descriptions.

Furthermore, the aggregator also has to be able to deal with a *discrepancy* in the knowledge about an entity available in different sources (*knowledge heterogeneity*): Due to different reasons (e.g., perspective, targeted applications), two sources might know more, less or different things about an entity. Exactly this makes their integration promising, but also challenging.

The aggregator should also be able to cope with the different *data quality levels* of the sources. User-generated datasets, or datasets generated by integrations of heterogeneous sources, are typically high in noise and missing values [2] which impedes entity identification. Finally, data brought together from multiple independently developed data sources that partially overlap, introduces *redundancy* and *conflicts* that need to be resolved in an efficient and consistent manner.

1.2 Approach and Contributions

To support entity-based searching over integrated entity descriptions of Web 2.0 data we propose an infrastructure for entity-based aggregation. Our main focus on the part of the infrastructure that matches queries to entity profiles. This requires extending search and aggregation technologies (e.g., semantic search engines, mashups, or portals) with functionallities that will allow them to maintain and exploit information about entities and their characteristics.

The current version of our system uses the Entity Name System (ENS) [23], which is an entity repository service. Our goal is to enable entity-based aggregation for the particular repository. This requires matching capabilities in order to ensure that each entity in the repository has a unique identifier. We achieve this through the search functionality. In short, when the repository is queried (i.e., entity search) we detect and return the entities from the repository that contain the characteristics found in the requested query. If no entity match is found in the repository, then a new entry is created in order to be used for future requests. If a partially matching entity is found, the stored entity may be enhanced with the additional characteristics from the source data, improving the chances of successful identification in future requests.

The introduced Matching Framework has several contributions regarding matching and entity searching. First, it brings data matching from its static and design-time nature into a dynamic process that starts at design-time with some initial knowledge and continues to evolve throughout its run-time use. An additional contribution is that it is able to employee an extensible set of matching methodologies, each based on a different matching technique. This allows us to deal with the inability of a single matching algorithm to address the matching problem. Also, matching is performed as a series of steps that include selection of the most appropriate matching algorithm and combination of results from more than one modules. Note that a short description of the Matching Framework was included in [19]. In this journal we provide the details of our work.

The remaining paper is structured as follows. Section 2 presents a motivating example. Section 3 introduces the Matching Framework and discusses the components composing it. Section 4 explains how existing matching methodologies can be incorporated in the proposed Matching Framework and also provides a couple of concrete examples. Section 5 describes various requests we used for evaluating our infrastructure and reports quality and execution time. Finally, Sect. 6 provides conclusions and discusses possible future directions.

2   Motivating Example

We are currently seeing a plethora of systems that use data from Web 2.0 applications and sources. For being able to use such data, the systems need to integrate the collected/received Web 2.0 data. Our matching framework focuses on providing this task, i.e., performing the integration of data coming from Web 2.0 applications and sources, and in particular on matching and merging together the data describing the same real world objects. Consider a portal system, such as Thoora[1], or Daylife[2], that crawls the web collecting, analysing, and categorizing news articles and blog posts. The system employs special similarity search techniques that find whether different articles talk about the same topic. These techniques are mainly based on textual analysis which try to identify entities mentioned in each article/post. Consider an article that talks about a person called Barack Obama living in Washington DC and a second one talking about a Nobel Prize winner also called Barack Obama who has been a senator in Illinois (i.e., entities labeled e_1 and e_2 in Fig. 1 respectively). The fact that both articles talk about a person called Barack Obama, is some indication that they may refer to the same person. However, it is not sure that they actually do, since the rest of the information they have is different. For instance, the second article, in contrast to the first, does not mention anything about the residence of the person, but instead mentions some origin place that is different from the Washington DC area.

A linked-data could link the two entities if there was a way to refer to them, for instance through the existence of some identifiers in the data, if such identifiers exist. Since the data is coming from text articles and not from database systems, this information may not be available at all, or may not be useful since these identifiers may have been locally assigned at run time with no persistence guarantees. The identification needs to be based mainly on their characteristics that are mentioned in the articles. Nevertheless, the portal developer, having the knowledge that Barack Obama is one of the Nobel Prize recipients can provide an add-hoc solution that links these two entities.

Consider now a blog post that talks about a person called Barack Obama, who is the recipient of a Nobel Prize and lives in Washington DC (shown as entity e_3 in Fig. 1). Clearly no document from those seen so far had an entity with these three characteristics. There are strong indications, however, that this

---

[1] http://www.thoora.com/.
[2] http://www.daylife.com/.

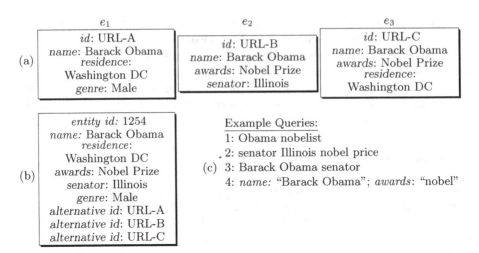

Fig. 1. (a) Three entities found in news articles and blog posts. (b) The corresponding entity entry created through entity-based aggregation. (c) A few query examples.

person is the same as the one in the first article, since they agree on the name and the residence place. It similarly agrees with the entity in the second article. Nevertheless, the portal developer, that knows that the entities in the first two articles are actually the same person, that has the combined set of their characteristics, can more confidently conclude that the entity mentioned in the blog post is the one in the first two. This kind of knowledge, however, resides in the mind of the portal developer alone with no systematic way of representing and using it. Linked-data, when available, may be exploited to reach similar conclusions, but it will require global knowledge, advanced reasoning and is not guaranteed to work in all cases.

To overcome this problem, during integration, the portal can utilize a repository of entities that contains discriminative knowledge about a large number of already known real world objects. These entities could have been collected from a number of existing social and semantic applications, such as Wikipedia, DBPedia, and Freebase, as well as articles that the portal crawler has already crawled and analyzed. The repository need not be a knowledge base, since this is neither realistic, nor practical. Instead, it should contain a collection of entities alongside their characteristics that allow their identification. Each entity, should also have a unique identifier that distinguishes it from all the other entities. The creation and use of such a repository requires solutions to a number of challenging issues. These issues include among others the efficient and effective entity search, population, storage, and maintenance of entities.

At run time, when the portal analyses an article and detects an entity, it collects its characteristics, and generated entity search queries using these characteristics. The repository receives this query and retrieves the identifier of the corresponding matched entity. Naturally, it may be the case that no entity with all these characteristics exists, or the case that there are more than one such entities. Thus, the repository responds to such a query request by providing a small list of candidates.

The richer the repository is, the more the confidence for the returned entities of a search request will be. If no entities are returned, then a new entry may be created in the repository for that new entity with the characteristics mentioned in the query, for example as shown in Fig. 1(b). This will enable its future identification in other articles or posts. If, instead, the entity is found in the repository through some partial match, then the characteristics of the found entity may be enhanced with the additional characteristics mentioned in the query to improve the chances and confidence of its identification in future articles.

3   Matching Framework

Consider again the challenges of integrating Web 2.0 data. Such data do not have a fixed schema and may consist of name-value attributes, attributes with only values, or a mix of both. Furthermore, the data given for integration may contain only one entity as well as a collection of entities which are somehow related to each other, such as entity repositories from Social applications. Despite the many existing results in entity matching, no solution has been found to work for all situations.

Our matching framework focuses on performing integration of data coming from Web 2.0 applications and sources, and in particular on matching and merging together the data describing the same real world objects. As such, it is responsible for receiving the queries requesting entities, controlling the execution flow, which primarily invokes the matching process, and returning the result set. Thus, our infrastructure can be useful for applications and systems that incorporate Web 2.0 data, for example the one described in the motivating example (i.e., Sect. 2).

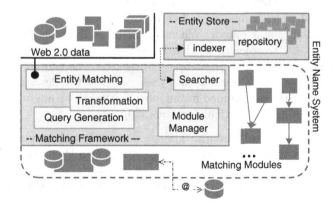

Fig. 2. An illustration of the matching framework within the ENS service.

Entity Name Service (ENS). The main components of the Matching Framework are illustrated in Fig. 2. Note that our framework is part of the Entity Name Service (ENS) [23]; an entity repository service that maintains a collection of entities as well as a number of discriminative properties that are used for entity identification and for assigning a unique identifier to each of the entities.

To achieve its goals, ENS addresses a set of challenges. The first challenge is to introduce a generic methodology for matching that incorporates an extendable set of matching modules, each focusing on addressing a specific matching situation. Another related challenge involves the selection of either the most promising module for a given query, or the combination of different modules. Additionally, the ENS service need to efficiently store and retrieve entities from the potentially very large entity collection and this is performed through the Entity Store.

Entity Store. A matching algorithm typically performs a lot of query-entity comparisons, which means that existing techniques either fail at Web scale or have performance that is prohibited for run-time. To minimize the time required for comparisons, we try to reduce the number of entities given to the matching algorithms for processing. The entity store provides a repository of entities along with an index for efficient entity retrieval. The requirement for the entity store is therefore to provide the set of entities that contain some of the information from the given query. By further processing these entities we can then identify the one requested by the query. We refer to the set of entities returned by the store as the *candidate set*.

In our current implementation we used NECESSITY system [20] as the entity store. NECESSITY is composed of two main part, which are shown in Fig. 2. The first part is the *repository*, implemented as a key-value Voldemort repository[3] and able to maintain a large number of entities. Scalability in Voldemort is obtained by simply adding more servers when more entities are added to the repository, which means that the remaining components of the ENS are left untouched. Moreover, this key-value repository supports linear scalability since it is designed from scratch to scale by simply adding new nodes. The second part of is the inverted *index*, implemented as a Solr Brocker[4], which uses Lucene[5] for full-text indexing and search. When a new query arrives, the Solr broker will assign the query to some of its shards, collect their *top-k* results, and then aggregate them to identify the best *top-k* entities, i.e., the candidate set that is then given to the matching framework. Please not that the details with respect to the entity store are not included in the journal since these are available in the publication describing the particular system [20].

Overview of Query Answering. The processing flow is controlled by the *Entity Matching* component, which also receives queries from users. When this component receives a query, it first gives it to the *Query Generation* and *Transformation* components for generating initial request commands for the

---

[3] http://project-voldemort.com/.
[4] http://lucene.apache.org/solr/.
[5] http://lucene.apache.org/.

Entity Store, and also to the *Module Manager* component for identifying the most suitable matching module or modules that could process this query. The initial request commands are then revised by the selected matching module(s) and then given to the *Searcher* component for pass it on to the Entity Store. The store processes the request commands and returns the candidate set. The entities from the candidate set are then given to the module(s) for performing matching and identifying the entity that corresponds to the given query. Through the *Entity Matching* component, the users receives the results for the processed queries.

The following paragraphs, we discuss the components composing the Matching Framework. We also give the details of the processing performing in each of these components.

3.1   Query Generation and Transformation

The Matching Framework needs to generate the request commands for the Entity Store. Our current system incorporates the NECESSITY entity store, thus, we need to generate a Lucene query. Since the Entity Store offers very efficient but restricted search functionality, this step might also require the generation of more that one queries, with the final candidate set being the merging of the results returned by the entity store for all generated queries.

We already explained, the query can be enhanced and refined by the matching modules according to their needs. This might involve query transformations on the schema level, for example to adapt from attributes used by the user/applications to attributes available in the repository, to include attribute alternatives, or to relax the query to the most frequent naming variants.

3.2   Matching Modules

Individual matching modules implement their own method for matching queries with candidates, i.e., entities returned from the store. Naturally, the algorithm of each module will focus on a specific matching task. For example, we can have matching modules providing algorithms for entities that do not contain attribute names, or for entities that contain inner-relationships. As shown in Fig. 2, modules may also use a local database for storing their internal data, or even communicate with external sources for retrieving information useful for their matching algorithm.

In addition to the individual modules, the Matching Framework can also contain modules that do not compute matches directly, but by combining the results of other modules. The current version of our system can handle the following two types of combination modules: (i) sequential and (ii) parallel processing. In sequential processing, the matching module invokes other modules in a sequence, and each module receives the results of previously invoked module. Therefore, each module refines the entity matches they receive and the resulted entity matches are the ones returned by the last module. The parallel processing invokes a set of matching modules at the same time. Each module returns its entity matches, and thus the combination module needs to combine their results.

3.3 Module Manager

This component is responsible for managing the modules included in the Matching Framework. To know the abilities of each module, the *Module Manager* maintains the module profiles. These profiles contain not only the module description and classification, but also information on their matching capabilities. For example, the average time required for processing queries and the query formats that they can handle.

The manager is responsible to select the best suitable matching module to perform the entity matching for the given entity. The *basic methodologies* for the performing the module selection are the following:

- The entity request explicitly defines the matching module that should be used. Such a selection can, for example, be based on previous experience or by knowing that a specialized matching module is more effective when integrating the data of a specific Social application.
- The matching module is selected based on information in the entities to be integrated. This may include requirements with respect to performance or supported entity types.
- The module is selected based on an analysis of the data in the entity to be integrated, for example existence or not of attribute names.

In addition to the *basic methodologies*, we can also have *advance methodologies*. Given the architecture of our Matching Framework we can actually perform the matching using more than one modules. This will result in a number of possible linkages, each one encoding a possible match between two entities along with the corresponding probability (indicating the belief we have for the particular match). Thus, the goal now is to maintain these linkages in the system and efficiently use them when processing incoming requests.

This is an interesting and promising direction that has not yet been studied deeply by the community. With respect to our Matching Framework, we have investigated how to perform only the required merges at run-time by following the possible world semantics given the linkages as well as effectively taking into consideration the entity specifications included in given requests [17]. In addition, we have also studied the possibility of executing complex queries instead of just answering entity requests. More specifically, we proposed the usage of an entity-join operator that allows expressing complex aggregation and iceberg/top-k queries over joins between such linkages with other data (e.g., relational tables) [16]. This includes a novel indexing structure that allows us to efficiently access the entity resolution information and novel techniques for efficiently evaluating complex probabilistic queries that retrieve analytical and summarized information over a (potentially, huge) collection of possible worlds.

4   Matching Modules

In this section we explain how the suggested system can incorporate and use existing matching techniques. For this task, it is important to understand the methodologies followed by the existing techniques. Thus, we consider the existing techniques categorized according to their methodology and explain how each category can be incorporated in our system.

Please note that we discuss existing matching techniques to the extend needed for this purpose of the introduced system. A complete overview of existing matching techniques as well as related surveys is available in [9,12,14,15,29].

4.1 Atomic Similarity Techniques

The first category of matching techniques consider that each entity is a single word or a small sequence of words. Matching issues is the entities of this category can occur from misspellings, abbreviations, acronyms, naming variants, and use of different languages (i.e., multilingualism). As examples consider the following: entity-1 is "TCCI Journal" and entity-2 is "Transactions on Computational Collective Intelligence journal". The matching methodology followed by the techniques of this category is based on detecting resemblance between the text values on the entities; more details about the followed methodology can be found in [6,8].

Module Example - Handing Multilingualism. In the modern global environments, the data is collected from many physically distributed sources or applications that may be located in different countries. This unavoidable leads to data expressed in a diverse set of languages. Furthermore, users with different cultural backgrounds typically pose queries in their native languages, that is not necessarily the same as the one in which the data has been stored. In such a situation, it is not at all surprising that traditional string matching algorithms are dramatically failing. A popular solution to cope with the problem is to use some form of standardization. For instance, as a convention, all the information can be translated into English and then stored into the system. When a query is received, it is first translated into English, if not already, and then evaluated against the stored data.

The specific approach has unfortunately two main limitations. The first is that the data will be returned to the users in English which will be surprising if the user knew that the data had been inserted is some other language. The second limitation is that the translation from one language to another is not always perfect. Thus, if the data was original in a language, say Italian, and the query posed by the user is also in Italian, translating both of them in English and then evaluating the query against the data may loose in terms of accuracy and precision.

For the above reasons, we follow an approach in which we combine the best of both worlds. In particular, we maintain two copies of the data. One that is the original form in which the data has been inserted. The second one, called the canonical representation, is the translation of the original data into English. When a user query is received, then two evaluations are taking place in parallel. The first is the evaluation of the user query as-is over the original data. This evaluation has the advantage that it allows the high-accuracy retrieval of the data that match the provided query in the same language. At the same time, the provided query is translated into the canonical form, i.e., English, and evaluated against the canonical forms of the data. This will allow the query to retrieve as answers, data that has been inserted originally in some completely different language. At the end, the result lists of the two approaches are merged together into one single list, by considering the combined returned score.

Another advantage of the approach we have followed, apart from the fact that it is easily extended to support more than one languages. Also, the translation mechanism allows for the utilization of additional information that guides the translation such as personalization, time, etc.

4.2 Entities as Sets of Data

This entity representation of this category can be seen as an extension of the previous one. More specifically, the entities are now represented as a small collection of data. The most typical situation is considering that each record of a relational table provides an entity. As examples consider the following: entity-1 is { "TCCI Journal", "2015", ...} and entity-2 is { "Transactions on Computational Collective Intelligence journal", "2015", ...}

One methodology to address this issue, it to concatenate the data composing each entity into one string and then perform matching using a technique from the previous category [7,21]. Other techniques perform matching by detecting mappings between entities. For example, [30] detects mappings by executing *transformations*, such as abbreviation, stemming, and the use of initials. Doan et al. [10] apply *profilers* that correspond to predefined rules with knowledge about specific concepts.

In the following paragraphs we present two such modules that have been included in our system. The first implements the methodology suggested from an existing technique, whereas the second is a new technique created for targeting a specific entity type.

Module Example II - Group Linkage. This module adapts the algorithm suggested in [24], and matches an entity to a candidate when it detects a large fraction of similarities between their data. For being able to use this algorithm we consider the given query Q and the candidate C as the two entities. Their matching probability is given by:

$$MP(Q,C) = \frac{\sum_{\forall p_i \in Q \forall p_j \in C} \begin{cases} sim(p_i, p_j) \ if \ sim > t \\ 0 \ otherwise \end{cases}}{|Q| + |C| - matched\_pairs,} \quad (1)$$

where $|C|$ gives the number of attribute value pairs in the candidate, $|Q|$ the number of predicates in the query, and $matched\_pairs$ the number of $sim(p_i, p_j)$ higher that threshold t.

Module Example II - Eureka Matching. The matching methodology of the Eureka module computes the overlap of the predicates in the query with the attributes of the candidates. As an initialization step, the algorithm creates a small local inverted index as follows: Each term (i.e., word) in the values from the attribute-value pairs of the query become keys in a hash table. We then process the information in each candidate and when we identify a candidate contains one of these values, we add the candidate's identifier with the values attribute to the list of entities of the corresponding key. The score $MP(Q, C)$ between the entity described in the query Q and candidate C is computed by:

$$\sum_{\forall p_1 \in Q, \forall p_2 \in C} \begin{cases} 1 \times importance(p_1.attr), \\ \quad if \ p_1.attr = p_2.attr \ \& \ p_1.value \in p_2.value \\ 0.5 \times importance(p_1.attr), \\ \quad if \ p_1.attr = null \ \& \ p_1.value \in p_2.value \end{cases} \quad (2)$$

where *importance* is a weight that reflects the importance of a specific attribute, e.g., attribute *name* is more important than attribute *residence* for the entities of Fig. 1.

4.3 Collective Matching

The techniques performing collective matching are based not only on the data composing each entity (as in Sects. 4.1 and 4.2) but also on available relationships and associations between the entities. For an example, consider that we are now working on addressing the entity matching problem in a collection of publications. One of the publications has authors α, β, and γ. Another publication has authors α, β, and γ'. Performing matching using one of the methodologies from the previous categories would result in the matching of authors α and β between the two publications, and a strong similarity between author γ and γ'. The former (i.e., matching of authors α and β) gives a relationship between the two publications. Combining this relationships with the similarity between γ and γ' increases our belief that these authors are actually the same, and thus, perform their match.

One approach for performing collection matching was introduced by Ananthakrishna et al. [4]. It detects fuzzy duplicates in dimensional tables by exploiting dimensional hierarchies. The hierarchies are built by following the links between the data from one table to the data from other tables. Entities are matched when the information across the generated hierarchies is found similar. The approach in [5] uses a graph structure, where nodes encode the entities and edges the inner-relationships between entities. The technique uses the edges from the graphs to cluster the nodes. The entities inside clusters are considered as matches and are then merged together. Another well-know algorithm is the *Reference Reconciliation* [11]. Matching starts by comparing the entity literals to detect possible relationships between entities. The detected matches are then propagated to the rest of the entities and is used for increasing the confident for the related entity matches. The approach introduced in [18] models the data into a Bayesian network, and uses probabilistic inference for computing the probabilities of entity matches and for propagating the information between matches.

Incorporating collective matching techniques is also possible with our system. The main required functionality for having such a module is being able to maintain the information related to the relationships between entities, including efficient update and navigation mechanism. This is a functionality that can be achieved through the entity store by storing relationships inside each entity, i.e., as attribute value pairs. The entity store can also implement specialized indexes for the relationships - if this is useful for the particular technique. Another functionality that would be helpful is being able to call other matching modules, and in particular modules from Sects. 4.1 and 4.2 for detecting possible similarities between entities. As we explained in Sect. 3 and also illustrate in Fig. 2, this capability can by realized through the matching framework thourgh sequential or parallel processing of modules.

4.4   Matching Using Schema Information

Another helpful source of knowledge related to entity matching is through the available schema information. Note that knowledge coming from schema information is typically correspondences between the entity attributes and not between the actual entities [13, 28], and thus can not directly be used for matching. However, it definitely assist the matching methodologies presented in the previous categories. For example, by knowing which schema attributes are identical, or present a high similarity, we perform a focused entity matching only on the particular attributes.

Our system can easily incorporate such methods. The most prominent mechanism is to implement them as a typical matching module. Schema information is anyway present in the entities of the Entity Store, and thus, accessible to the modules. The entity matching process that uses the results of this processing can be included in the same module, or in another module that just call it.

4.5   Blocking-Based Matching

One methodology to increase the efficiency of matching is by reducing the performed entity comparisons. Blocking is focusing on this, and more specifically through the following process: entities are separated into blocks, such that the entities of the same block can result in one or more matches, whereas the entities of different blocks can not result in matches. Having such blocks means that we do not need to compare all entities between them but only the entities inside the same block, which of course reduces the comparisons.

The most common methodology for blocking is to associate each entity with a value summarizing key its data and use this to create the blocks. For example, the approach introduced in [22], builds high-dimensional overlapping blocks using string similarity metrics. Similarly, the approach in [1], uses suffixes of the entity data. One group of approaches focused on data with heterogeneous semi-structured shemata. For instance, [25, 26] introduced an attribute-agnostic mechanism for generating the blocks and explained how efficiency can be improved through scheduling the order of block processing and identifying when to stop the processing. The approach introduced in [31] processes iteratively blocks in order to use the results of one block in the processing step of another block. The idea of iteratively block processing was also studied in [27]. It provided a principled framework with message passing algorithms for generating a global solution for the entity resolution over the complete collection.

Incorporating blocking-based matching in our system can be achieved with two ways. The first is by modifying the storage functionality, and in particular, the method for adding them in the Entity Store. Entities should be also processed to generate the value summarizing key which is also included (and stored) in their attribute value pairs. These keys are used by the matching modules for skipping entity comparisons. The alternative incorporation of blocking is by modifying the matching framework in order to follow one specific blocking technique, i.e., entities are actually separated using the value summarizing key and are given to the modules grouped according to the block in which they belong.

5   Usage Experience

We now demonstrate and discuss the applicability of the suggested infrastructure. For this we use various collection of requests that can be used to evaluate methodologies for entity linkage as well as entity search. Each collection is taken from a different real world data source (e.g., structured and unstructured data) and this leads to requests that have different format. Note that the entity requests are accompanied with a small set of urls from the systems in which they are described (e.g., Wikipedia url, OKKAM id, etc.). The following paragraphs describe these entity collections[6] and report the required processing time as well as the quality of the returned answer set.

(A) People Information. The first collection contains 7444 entity requests from short descriptions of people's data from Wikipedia. The text describing these people was crawled and then processed using the OpenCalais extractor to extract the contained entities. Some examples are: (i) "Evelyn Dubrow" Country="US" Position="womenlabor advocate" and (ii) "Chris Harman" Position="activist" Position="journalist".

(B) News Events & Web Blog. Wikipedia contains small summaries of news events reported in various online systems, such as BBC and Reuters. We used the OpenCalais extractor to identify the entities in the events. This also resulted in entity type along a few name-value attributes for each entity. We also used the OpenCalais extractor to identify the entities described in a small set of blogs discussing political events, e.g., http://www.barackoblogger.com/. The process resulted in a collection with 1509 entity requests. Some request examples are: (i) Alex Rodriguez and (ii) name="Charles Krauthammer".

(C) Historical Records. Web pages sometimes contain local lists of entities, for example members of organizations, or historic records in online newspapers. This collection contains 183 entity requests taken from such lists, i.e., no extraction

---

[6] The collections can be found at http://www.softnet.tuc.gr/~ioannou/entityrequests.
html.

process was involved. Some examples are: (i) Don Henderson British actor and (ii) George Wald American scientist recipient of the Nobel Prize in Physiology or Medicine.

(D) DBPedia. The last entity collection contains 2299 requests from the structured DBPedia data. Some examples are: (i) name="Jessica Di Cicco" occupation="Actress Voice actress" yearsactive="1989-present" and (ii) birth-name="Jessica Dereschuk" eyeColor="Green" ethnicity="WhiteCaucasian" hairColor="Blonde" years="2004".

To evaluate the introduced infrastructure we used the NECESSITY entity store [20] with ~6.8 million entities. The entities in the store were people and organizations from Wikipedia, and geographical items from GeoNames. We then used the infrastructure to retrieve the entities for 4000 requests from our four collections. For each request the infrastructure returned a list of entities matching the particular request. The requested entity was the first item in the list for 77.5 %. The average time for processing the requests was 0.025 s.

6   Conclusions

In this work we presented an novel approach for enabling aggregators to perform entity-based integration, leading to more efficient and effective integration of Social and Semantic Web data. In particular we equipped aggregators with an Entity-Name-System, which offers storage and matching functionality for entities. The matching functionality is based on a generic framework that allows incorporation and combination of an expandable set of matching modules. The results of an extended experimental evaluation on Web data, such as Web blogs and news articles, demonstrated the efficiency and effectiveness of our approach.

Ongoing work includes the improvement of the existing functionality based on the fact that many entity integration requests may arrive at the same time to the aggregator. This requires a special handling both for reasons of efficiency and for reasons of accuracy. We refer to this task as *bulk integration.* The entities given for bulk integration may exhibit some special characteristics, such as presence of inner-relationships, which should be also considered to improve integration's performance and quality. We are also investigating the ways of improving integration by exploiting external knowledge in systems, such as WordNet. Another interesting direction, which we are currently investigating, is how to *combine* and reason over the matching results that have been generated by more than one matching modules.

References

1. Aizawa, A., Oyama, K.: A fast linkage detection scheme for multi-source information integration. In: WIRI, pp. 30–39 (2005)
2. Alexe, B., Tan, W.C., Velegrakis, Y.: STBenchmark: towards a benchmark for mapping systems. PVLDB 1(1), 230–244 (2008)

3. Amer-Yahia, S., Markl, V., Halevy, A.Y., Doan, A., Alonso, G., Kossmann, D., Weikum, G.: Databases and Web 2.0 panel at VLDB 2007. SIGMOD Rec. **37**, 49–52 (2008)
4. Ananthakrishna, R., Chaudhuri, S., Ganti, V.: Eliminating fuzzy duplicates in data warehouses. In: VLDB (2002)
5. Bhattacharya, I., Getoor, L.: Deduplication and group detection using links. In: LinkKDD (2004)
6. Bilenko, M., Mooney, R., Cohen, W., Ravikumar, P., Fienberg, S.: Adaptive name matching in information integration. IEEE Intell. Syst. **18**(5), 16–23 (2003)
7. Cohen, W.: Data integration using similarity joins and a word-based information representation language. TOIS **18**(3), 288–321 (2000)
8. Cohen, W., Ravikumar, P., Fienberg, S.: A comparison of string distance metrics for name-matching tasks. In: IIWeb Co-located with IJCAI, pp. 73–78 (2003)
9. Doan, A., Halevy, A.Y.: Semantic integration research in the database community: a brief survey. AI Mag. **26**, 83–94 (2005)
10. Doan, A., Lu, Y., Lee, Y., Han, J.: Object matching for information integration: a profiler-based approach. In: IIWeb Co-located with IJCAI, pp. 53–58 (2003)
11. Dong, X., Halevy, A., Madhavan, J.: Reference reconciliation in complex information spaces. In: SIGMOD Conference, pp. 85–96 (2005)
12. Elmagarmid, A.K., Ipeirotis, P.G., Verykios, V.S.: Duplicate record detection: a survey. TKDE **19**, 1–16 (2007)
13. Fagin, R., Haas, L.M., Hernández, M., Miller, R.J., Popa, L., Velegrakis, Y.: Clio: schema mapping creation and data exchange. In: Borgida, A.T., Chaudhri, V.K., Giorgini, P., Yu, E.S. (eds.) Conceptual Modeling: Foundations and Applications. LNCS, vol. 5600, pp. 198–236. Springer, Heidelberg (2009)
14. Ferrara, A., Nikolov, A., Scharffe, F.: Data linking for the semantic web. J. Data Semant. **7**(3), 46–76 (2011)
15. Getoor, L., Diehl, C.P.: Link mining: a survey. SIGKDD Explor. **7**, 3–12 (2005)
16. Ioannou, E., Garofalakis, M.: Query analytics over probabilistic databases with unmerged duplicates. TKDE **27**(8), 2245–2260 (2015)
17. Ioannou, E., Nejdl, W., Niederée, C., Velegrakis, Y.: On-the-fly entity-aware query processing in the presence of linkage. PVLDB **3**(1), 429–438 (2010)
18. Ioannou, E., Niederée, C., Nejdl, W.: Probabilistic entity linkage for heterogeneous information spaces. In: Bellahsène, Z., Léonard, M. (eds.) CAiSE 2008. LNCS, vol. 5074, pp. 556–570. Springer, Heidelberg (2008)
19. Ioannou, E., Niederée, C., Velegrakis, Y.: Enabling entity-based aggregators for web 2.0 data. In: WWW, pp. 1119–1120 (2010)
20. Ioannou, E., Sathe, S., Bonvin, N., Jain, A., Bondalapati, S., Skobeltsyn, G., Niederée, C., Miklos, Z.: Entity search with NECESSITY. In: WebDB (2009)
21. Koudas, N., Marathe, A., Srivastava, D.: Flexible string matching against large databases in practice. In: VLDB, pp. 1078–1086 (2004)
22. McCallum, A., Nigam, K., Ungar, L.: Efficient clustering of high-dimensional data sets with application to reference matching. In: KDD, pp. 169–178 (2000)
23. Miklós, Z., et al.: From Web data to entities and back. In: Pernici, B. (ed.) CAiSE 2010. LNCS, vol. 6051, pp. 302–316. Springer, Heidelberg (2010)
24. On, B.W., Koudas, N., Lee, D., Srivastava, D.: Group linkage. In: ICDE (2007)
25. Papadakis, G., Ioannou, E., Niederée, C., Fankhauser, P.: Efficient entity resolution for large heterogeneous information spaces. In: WSDM, pp. 535–544 (2011)
26. Papadakis, G., Ioannou, E., Niederée, C., Palpanas, T., Nejdl, W.: Beyond 100 million entities: large-scale blocking-based resolution for heterogeneous data. In: WSDM, pp. 53–62 (2012)

27. Rastogi, V., Dalvi, N., Garofalakis, M.: Large-scale collective entity matching. PVLDB **4**(4), 208–218 (2011)
28. Shen, W., DeRose, P., Vu, L., Doan, A., Ramakrishnan, R.: Source-aware entity matching: a compositional approach. In: ICDE, pp. 196–205 (2007)
29. Staworko, S., Ioannou, E.: Management of inconsistencies in data integration. In: Data Exchange, Integration, and Streams, pp. 217–225 (2013)
30. Tejada, S., Knoblock, C.A., Minton, S.: Learning domain-independent string transformation weights for high accuracy object identification. In: KDD (2002)
31. Whang, S., Menestrina, D., Koutrika, G., Theobald, M., Garcia-Molina, H.: Entity resolution with iterative blocking. In: SIGMOD Conference, pp. 219–232 (2009)

Author Index

Printed in the United States
By Bookmasters